ST(P) MATHEMATICS 3B

ST(P) MATHEMATICS will be completed as follows:

Published 1984
- **ST(P) 1**
- **ST(P) 1** Teacher's and Answer Book
- **ST(P) 2**

Published 1985
- **ST(P) 2** Teacher's and Answer Book
- **ST(P) 3A**
- **ST(P) 3B**
- **ST(P) 3A** Teacher's and Answer Book
- **ST(P) 3B** Teacher's and Answer Book

In preparation
- **ST(P) 4A**
- **ST(P) 4B**
- **ST(P) 4A** Teacher's and Answer Book
- **ST(P) 4B** Teacher's and Answer Book
- **ST(P) 5A**
- **ST(P) 5B**
- **ST(P) 5A** Teacher's and Answer Book
- **ST(P) 5B** Teacher's and Answer Book

ST(P) MATHEMATICS 3B

L. Bostock, B.Sc.
formerly Senior Mathematics Lecturer, Southgate Technical College

S. Chandler, B.Sc.
formerly of the Godolphin and Latymer School

A. Shepherd, B.Sc.
Head of Mathematics, Redland High School for Girls

E. Smith, M.Sc.
Head of Mathematics, Tredegar Comprehensive School

Stanley Thornes (Publishers) Ltd

© L. Bostock, S. Chandler, A. Shepherd, E. Smith, 1985

All rights reserved. No part of this publication may be reproduced, stored in a retrieval system or transmitted in any form or by any means, electronic, mechanical, photocopying, recording or otherwise, without the prior written consent of the copyright holders. Applications for such permission should be addressed to the publishers: Stanley Thornes (Publishers) Ltd, Old Station Drive, Leckhampton, CHELTENHAM GL53 0DN.

First Published 1985 by
Stanley Thornes (Publishers) Ltd,
Old Station Drive,
Leckhampton,
CHELTENHAM GL53 0DN

Reprinted 1986

British Library Cataloguing in Publication Data
ST(P) mathematics
  Book 3B
  1. Mathematics—1961–
  I. Bostock, L.
  510 QA39.2

ISBN 0-85950-209-0

Typeset by Eyre & Spottiswoode Ltd, Thanet Press, Margate, Kent
Printed and bound in Great Britain by Ebenezer Baylis and Son, Worcester

CONTENTS

Introduction   **viii**

Chapter 1   **Basic Arithmetic**   **1**
Prime numbers. Factors. Prime factors. Multiples. Squares. Multiplying a string of numbers. Index notation. Mixed operations of $+$, $-$, $\times$ and $\div$.

Chapter 2   **Fractions and Decimals**   **9**
Equivalent fractions. Simplifying fractions. Mixed numbers. Addition, subtraction, multiplication and division of fractions. Reciprocals. Interchanging decimals and fractions. Addition, subtraction, multiplication and division of decimals. Decimal places.

Chapter 3   **Symmetry and Reflections**   **29**
Line symmetry. Reflections. Mirror lines.

Chapter 4   **Rotations**   **38**
Rotational symmetry. Rotations. Centre of rotation. Finding an angle of rotation. Finding an image.

Chapter 5   **Directed Numbers**   **46**
The number line. Addition, subtraction, multiplication and division of directed numbers.

Chapter 6   **Brackets and Equations**   **51**
Collecting like terms. Multiplying. Expansion and collection of like terms with brackets. Linear equations. Common factors.

Chapter 7   **Coordinates and Straight Line Graphs**   **61**
Coordinates. Lines parallel to the axes. Slant lines through the origin. Gradients. The equation $y = mx + c$. Intersection.

Chapter 8   **Translations**   **95**
Vectors. Translations. Transformations.

Chapter 9   **Essential Geometry**   **106**
Revision of the basic facts about angles, triangles and parallel lines.

Chapter 10   **Percentages**   **122**
Percentages as fractions and decimals. Percentages of quantities. Fractions to percentages. One quantity as a percentage of another.

Chapter 11 **Matrices** 135
Matrices as stores of information. Size of a matrix. Addition and subtraction of matrices. Multiples of matrices.

Chapter 12 **Averages. Speed** 153
Average of a set of quantities. Speed, distance and time and the relationship between them. Average speed.

Chapter 13 **Polygons and Tessellations** 164
Regular polygons and rotational symmetry. Drawing regular polygons. Exterior angles. Interior angles. Tessellations.

Chapter 14 **Formulae** 180
Substituting numbers into a formula. Changing the subject of a formula.

Chapter 15 **Simple Interest** 197
Using the formula $I = PRT/100$

Chapter 16 **Angles in a Circle** 201
Minor and major arcs. Sectors and segments. Angle subtended by an arc. Angles on the same arc. Angle at the centre and angle at the circumference on the same arc. Angle in a semicircle.

Chapter 17 **Inequalities** 219
The symbols $>$ and $<$. Illustration on a number line. Addition and subtraction involving inequalities. Solving an inequality. Multiplication and division by a positive number.

Chapter 18 **Statistics** 229
Making a frequency table from a bar chart and making a bar chart from a frequency table. Making a frequency table from unsorted information. Grouping information. Reading pie charts. Drawing pie charts. Pictographs.

Chapter 19 **Cyclic Quadrilaterals** 246
Opposite angles in a cyclic quadrilateral. Exterior angle property of cyclic quadrilaterals.

Chapter 20 **Ratio and Proportion** 255
Comparing sizes. Simplifying ratios. Expressing ratios in the form $1:n$ and $n:1$. Comparing three quantities. Division in a given ratio. Map scales. Direct proportion.

Chapter 21 Indices and Approximations   274
Addition and subtraction rules. Zero and negative indices. Standard form. Calculators and scientific notation. Significant figures. Rough approximations. Using a calculator for multiplication and division.

Chapter 22 The Tangent of an Angle   284
Naming the sides of a right-angled triangle. The tangent of an angle. Finding the tangent of an angle from a right-angled triangle. Using a calculator. Finding sides and angles in a right-angled triangle.

Chapter 23 The Sine and Cosine of an Angle   303
The sine of an angle. Using a calculator. Finding sides and angles in a right-angled triangle. The cosine of an angle. Using a calculator. Finding sides and angles in a right-angled triangle.

Chapter 24 Algebraic Products   320
Product of two brackets. Finding a pattern and using the pattern. Perfect squares and the difference between two squares.

Chapter 25 Mean, Mode and Median   333
Finding the mean, mode and median from a set of figures. Finding the mean, mode and median from a frequency table and a bar chart.

Chapter 26 Pythagoras' Theorem   348
Squares and square roots from a calculator. Pythagoras' Theorem and using it to find a side in a right-angled triangle.

Chapter 27 Areas   364
Revision of basic units of length and area. Area of a rectangle, triangle, parallelogram and trapezium. Compound shapes.

Chapter 28 Circle Calculations   388
The circumference and area of a circle.

Chapter 29 Percentage Problems   401
Profit and loss. VAT. Discount. Finding the original quantity.

Multiple Choice Revision Exercises   413

INTRODUCTION

This book has been written for those of you who are aiming at the middle and lower level GCSE papers in Mathematics. The book contains a good deal of revision of earlier work which should help you to consolidate your skills with basic mathematics before you move on to new topics. You will probably find that you do not need to cover all the topics in this book as some may not be in the syllabus that your school uses. As is the case with the earlier books in the series there are plenty of straightforward questions and the exercises are divided into three types of question.

The first type, identified by plain numbers, e.g. **12.**, helps you to see if you understand the work. These questions are considered necessary for every chapter you attempt.

The second type, identified by a single underline, e.g. **12.**, are extra, but not harder, questions for quicker workers, for extra practice or for later revision.

The third type, identified by a double underline, e.g. **12.**, are for those of you who manage type 1 questions fairly easily and therefore need to attempt slightly harder questions.

Most chapters end with mixed exercises and with multiple choice questions. These will help you revise what you have done, either when you have finished the chapter or at a later date.

Finally a word of advice: when you arrive at an answer, whether with the help of a calculator or not, always ask yourself 'Is my answer a reasonable one for the question that was asked?'

1 BASIC ARITHMETIC

PRIME NUMBERS

A number that can be divided exactly only by itself and 1 is called a prime number; e.g., 7 is a prime number (as 7 can only be 7×1) but 6 is not a prime number (because $6 = 6 \times 1$ or 3×2).

The smallest prime number is 2, as 1 is *not* a prime number.

EXERCISE 1a
1. Write down the first four prime numbers, starting with 2.
2. Write down the prime numbers that are less than 10.
3. Write down the prime numbers that are between 10 and 20.
4. In the set $\{2, 3, 4, 5, 6, 7, 8\}$, which members are prime numbers?
5. In the set $\{5, 7, 9, 11, 13, 15\}$, which members are prime numbers?
6. In the set $\{12, 13, 14, 15, 16, 17\}$, which members are prime numbers?
7. In the set $\{20, 21, 22, 23, 24, 25\}$, which members are prime numbers?
8. Apart from 2, are there any even numbers that are prime numbers?

FACTORS

A factor of a number divides exactly into that number, e.g. 2 is a factor of 10 because 2 divides into 10 exactly 5 times.

On the other hand 3 is not a factor of 10 because 3 does not divide exactly into 10.

EXERCISE 1b
1. Is 2 a factor of
   a) 4   b) 9   c) 12   d) 21   e) 84 ?
2. How can you tell whether 2 is a factor of a number?
3. Is 5 a factor of
   a) 8   b) 10   c) 15   d) 25   e) 32   f) 85 ?
4. How can you tell whether 5 is a factor of a number?

5. Write down those members of the set { 8, 16, 40, 35, 41, 206, 515 } for which

   a) 2 is a factor   b) 5 is a factor.

6. Is 3 a factor of

   a) 6   b) 9   c) 10   d) 15   e) 16 ?

7. The number 3 is a factor of 51. Add the digits of 51.

   Is 3 a factor of your answer?

8. The numbers 75 and 102 are each divisible by 3.

   a) Add the digits of 75. Is your answer divisible by 3 ?
   b) Add the digits of 102. Is your answer divisible by 3 ?

From the last exercise we see that

> any even number is divisible by 2
>
> any number ending in 0 or 5 is divisible by 5
>
> when the sum of the digits of a number is divisible by 3, the number itself is divisible by 3.

EXERCISE 1c

> Write down all the factors of 12.
>
> 1, 2, 3, 4, 6, 12 are all factors of 12
>
> (Note that factors of a number are not always prime.)

Write down all the factors of:

1. 6
2. 8
3. 10
4. 4
5. 9
6. 18
7. 15
8. 16
9. 21
10. 26
11. 51
12. 19

PRIME FACTORS

Sometimes we need to know just the prime factors of a number and to be able to express the number as the product of its prime factors.

For example, the prime factors of 6 are 2 and 3

and $6 = 2 \times 3$

For larger numbers we need a more systematic approach.
Consider the number 84.

The lowest prime factor of 84 is 2,
so we divide 2 into 84 $\qquad 84 \div 2 = 42$

We then repeat the process with 42 $\qquad 42 \div 2 = 21$

We repeat the process again with 21
but 3 is the lowest prime factor of 21 $\qquad 21 \div 3 = 7$

Repeating the process again $\qquad 7 \div 7 = 1$

Therefore the prime factors of 84 are 2, 2, 3 and 7

i.e. $\qquad 84 = 2 \times 2 \times 3 \times 7$

It is easier to keep track of the prime factors if the division is set out like this:

$$\begin{array}{r} 2\,)\,\overline{84} \\ 2\,)\,\overline{42} \\ 3\,)\,\overline{21} \\ 7\,)\,\overline{7} \\ 1 \end{array}$$

EXERCISE 1d

Express 90 as the product of its prime factors.

$90 = 2 \times 3 \times 3 \times 5$

$$\begin{array}{r} 2\,)\,\overline{90} \\ 3\,)\,\overline{45} \\ 3\,)\,\overline{15} \\ 5\,)\,\overline{5} \\ 1 \end{array}$$

Express the following numbers as products of their prime factors.

1. 10   **4.** 12   **7.** 60   **10.** 66
2. 21   **5.** 8   **8.** 50   **11.** 126
3. 35   **6.** 28   **9.** 36   **12.** 108

MULTIPLES

If a number divides exactly into a second number, the second number is a multiple of the first.

For example, the first six multiples of 3 are   3, 6, 9, 12, 15, 18

EXERCISE 1e

1. From the set $\{2, 3, 5, 6, 8, 10, 12, 14, 15, 16, 18, 20\}$ write down the members that are

   a) multiples of 2
   b) multiples of 3
   c) multiples of 4
   d) multiples of 5
   e) multiples of 6
   f) multiples of 8

2. Write down the first four multiples of

   a) 7   b) 5   c) 8   d) 10   e) 12   f) 15

3. Write down the multiples of 9 between 50 and 100.

4. Write down the multiples of 7 between 10 and 50.

5. Write down the multiples of 11 between 10 and 100.

SQUARES

A square number, or a perfect square, can be written as the product of two equal numbers.

For example, 16 is a perfect square because $16 = 4 \times 4$.

EXERCISE 1f

1. Write down the perfect squares between 2 and 10.

2. Write down the perfect squares between 10 and 101.

3. In the set $\{2, 4, 6, 8, 10, 12, 16, 18, 20\}$, which members are perfect squares?

4. In the set $\{2, 4, 8, 16, 32, 64, 128\}$, which members are perfect squares?

5. In the set $\{3, 9, 27, 81\}$, which members are perfect squares?

6. A perfect cube is the product of three equal numbers, e.g. $27 = 3 \times 3 \times 3$.
   Find the prime factors of each number in the set $\{6, 8, 25, 81, 125\}$. Hence write down those members of the set that are perfect cubes.

Basic Arithmetic 5

MULTIPLYING A STRING OF NUMBERS

Consider $2 \times 3 \times 6$; this means multiply 2 by 3 and then multiply the result by 6,

therefore
$$2 \times 3 \times 6 = 6 \times 6$$
$$= 36$$

Similarly $\quad 2 \times 3 \times 4 \times 5 = 6 \times 4 \times 5 = 24 \times 5 = 120$

EXERCISE 1g Calculate:

1. $2 \times 4 \times 3$
2. $3 \times 2 \times 5$
3. $5 \times 4 \times 2$
4. $7 \times 2 \times 2$
5. $3 \times 5 \times 2$
6. $4 \times 5 \times 3$
7. $6 \times 2 \times 3$
8. $3 \times 7 \times 2$
9. $2 \times 5 \times 2 \times 3$
10. $4 \times 2 \times 4 \times 3$
11. $3 \times 7 \times 1 \times 5$
12. $5 \times 2 \times 3 \times 4$

INDEX NOTATION

In the number 2^3, the small 3 is called an index, or power, and 2^3 means "three 2s multiplied together".

i.e. $\quad 2^3 = 2 \times 2 \times 2$

so the value of 2^3 is 8.

In the same way we can use indices to write

$$4 \times 4 \times 4 \times 4 \times 4 \quad \text{as} \quad 4^5$$

We write 4^5 but we *say* "4 to the power 5" or "4 to the five".

Similarly we can read 3^2 as "3 to the power 2" although in this case we can also read 3^2 as "3 squared".

For 3^3 we say either "3 to the power 3" or "3 cubed".

EXERCISE 1h Find the value of:

1. 2^3
2. 3^2
3. 4^2
4. 5^2
5. 2^4
6. 3^3
7. 4^3
8. 6^2
9. 5^3
10. 3^4
11. 7^2
12. 2^5

> Calculate: a) $2^2 \times 3^4$ b) $(3^2)^3$
>
> a) $\quad 2^2 \times 3^4 = 2 \times 2 \times 3 \times 3 \times 3 \times 3$
> $ = 4 \times 81$
> $ = 324$
>
> b) $\quad (3^2)^3 = (3 \times 3)^3$
> $ = 9^3$
> $ = 9 \times 9 \times 9$
> $ = 81 \times 9$
> $ = 729$

Find the value of:

13. $3^2 \times 2^3$ **17.** $5^2 \times 2^2$ **21.** $2^3 \times 3$

14. $4^2 \times 2^2$ **18.** $7^2 \times 2$ **22.** 3×5^2

15. $(2^3)^2$ **19.** $(4^2)^2$ **23.** $(3^2)^2$

16. $6^2 \times 2^2$ **20.** 10^4 **24.** $7^2 \times 2^2$

MIXED OPERATIONS

When a calculation involves a mixture of addition, subtraction, multiplication and division, we do the multiplication and division first.

For example $\qquad 3 + 4 \times 2 = 3 + 8$

It helps if brackets are put round the parts that are to be done first.

EXERCISE 1i

> Calculate $6 \div 2 + 4 \times 2 - 1$
>
> $6 \div 2 + 4 \times 2 - 1 = (6 \div 2) + (4 \times 2) - 1$
> $ = 3 + 8 - 1$
> $ = 10$

Basic Arithmetic

Calculate:

1. $8 - 2 \times 2$
2. $3 \times 4 - 6$
3. $12 \div 4 + 3$
4. $5 + 2 \times 3$
5. $3 \times 7 - 8$
6. $10 \div 2 + 2$
7. $3 - 4 \div 2$
8. $8 + 2 \times 2$
9. $16 - 3 \times 2$
10. $5 \times 3 - 5$
11. $7 + 12 \div 2$
12. $24 \div 8 - 2$

13. $3 + 4 - 6 \div 3$
14. $2 \times 7 + 8 \div 4$
15. $3 + 2 \times 4 - 5$
16. $7 - 10 \div 2 + 3$
17. $5 \times 2 - 3 + 1$
18. $18 \div 3 + 6 \div 2$
19. $3 + 1 \times 3 - 2$
20. $12 \div 4 + 3 \times 2$
21. $6 + 3 \times 2 - 4$
22. $10 - 15 \div 3 + 2$
23. $7 \times 3 - 6 + 2$
24. $14 - 2 \times 4 + 7$

MIXED EXERCISES

EXERCISE 1j

1. From the set $\{1, 2, 5, 9, 12, 18, 21, 36, 39, 41\}$ write down the numbers which are

    a) prime
    b) odd
    c) even
    d) multiples of 2
    e) multiples of 3
    f) factors of 36

2. Find the highest number which is a factor of

    a) both 6 and 4
    b) both 15 and 20
    c) both 8 and 20

3. Find the lowest number that is a multiple of

    a) both 2 and 3
    b) both 4 and 5
    c) both 3 and 4

4. Calculate:

    a) $2 \times 4 \times 3$
    b) $2 + 4 \times 3$
    c) $2 \times 4 + 3$

5. Calculate:

    a) 3^3
    b) $(2^2)^2$

6. Calculate:

    a) $12 \div 6 + 4$
    b) $3 + 4 \times 2$
    c) $7 - 4 \div 2$

EXERCISE 1k

1. From the set $\{4, 5, 6, 8, 12, 15, 20, 21, 23, 27, 29\}$ write down the numbers which are

    a) odd

    b) multiples of 5

    c) factors of 40

    d) prime

    e) even

    f) multiples of 4

2. Calculate:

    a) $6 \times 2 \times 4$ b) $6 \times 2 - 4$ c) $16 - 2 \times 4$

3. Calculate:

    a) $12 - 5 \times 2$ b) $4 \div 2 + 3$ c) $3 \times 2 \times 4 \times 3$

4. Calculate:

    a) 2^4 b) $(3^2)^2$

EXERCISE 1l In this exercise you are given several alternative answers. Write down the letter that corresponds to the correct answer.

1. The prime numbers in the set $\{1, 2, 4, 7, 11, 15\}$ are

    A 1, 2, 7, 11   **B** 2, 4   **C** 2, 7, 11   **D** 2, 7, 11, 15

2. The value of $2 + 3 \times 2$ is

    A 7   **B** 10   **C** 6   **D** 8

3. The square numbers (perfect squares) in the set $\{2, 4, 8, 9, 10, 16, 20\}$ are

    A 4, 9, 16   **B** 4, 8, 10, 16, 20   **C** 2, 4, 8, 9, 16

4. The highest number which is a factor of both 4 and 6 is

    A 2   **B** 4   **C** 1   **D** 12

5. The value of $3^2 \times 2$ is

    A 36   **B** 18   **C** 25   **D** 12

6. The value of $3000 - 3$ is

    A 2007   **B** 2997   **C** 2907   **D** 2700

2 FRACTIONS AND DECIMALS

FRACTIONS

EXERCISE 2a Write down the fraction of the figure that is shaded.

1.

2.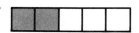

3.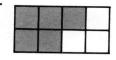

4.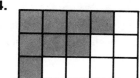

5.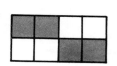

6.

7.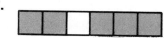

8.

9.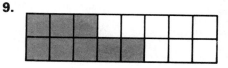

10.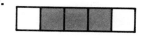

11. Copy the diagram on to squared paper and then shade $\frac{2}{3}$ of it. How many twelfths of your figure is this?

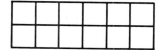

12. Copy the diagram on to squared paper and shade $\frac{1}{3}$ of it. How many fifteenths of your figure is this?

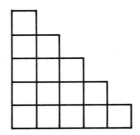

EQUIVALENT FRACTIONS

These diagrams show that $\frac{1}{2}$, $\frac{2}{4}$, $\frac{4}{8}$ and $\frac{3}{6}$ are *equivalent fractions*.

We can find fractions equivalent to a given fraction by multiplying the numerator *and* the denominator by the same number,

e.g. $\qquad \frac{1}{2} = \frac{1 \times 5}{2 \times 5} = \frac{5}{10}$

and $\qquad \frac{3}{4} = \frac{3 \times 2}{4 \times 2} = \frac{6}{8}$

EXERCISE 2b

Find the missing number to make $\frac{6}{\_}$ equivalent to $\frac{2}{3}$

$$\frac{2}{3} = \frac{6}{\_}$$

(If $\frac{2}{3} = \frac{6}{\_}$ the numerator has been multiplied by 3)

Therefore $\qquad \frac{2}{3} = \frac{2 \times 3}{3 \times 3} = \frac{6}{9}$

Fractions and Decimals 11

Fill in the missing number to make the fractions equivalent.

1. $\frac{1}{4} = \frac{}{8}$
2. $\frac{3}{4} = \frac{}{12}$
3. $\frac{1}{8} = \frac{2}{}$
4. $\frac{2}{3} = \frac{8}{}$
5. $\frac{2}{5} = \frac{}{10}$

6. $\frac{1}{3} = \frac{}{9}$
7. $\frac{3}{5} = \frac{9}{}$
8. $\frac{3}{7} = \frac{6}{}$
9. $\frac{5}{9} = \frac{}{27}$
10. $\frac{3}{8} = \frac{}{32}$

11. Express $\frac{2}{3}$ as ninths.
12. Express $\frac{3}{4}$ as sixteenths.
13. Express $\frac{3}{10}$ as thirtieths.
14. Express $\frac{1}{12}$ as thirtysixths.
15. Express $\frac{4}{5}$ as twentieths.
16. Express $\frac{5}{8}$ as twentyfourths.

SIMPLIFYING FRACTIONS

We know that $\frac{2}{4} = \frac{1}{2}$

When we write $\frac{2}{4}$ as $\frac{1}{2}$ we have *simplified* $\frac{2}{4}$.

When the numerator and denominator of a fraction have a common factor, we can simplify the fraction if we divide both numerator and denominator by that common factor.

Consider the fraction $\frac{9}{15}$.

The numbers 9 and 15 have a common factor of 3.

Dividing each by 3 gives $\frac{\cancel{9}^3}{\cancel{15}_5} = \frac{3}{5}$

The numbers 3 and 5 do not have any common factors so we cannot simplify the fraction any further.

When this is the case we say that the fraction is expressed in its *lowest terms* or in its simplest form.

EXERCISE 2c

Simplify $\frac{15}{45}$ as far as possible.

$$\frac{\cancel{15}^{1}}{\cancel{45}_{3}} = \frac{1}{3}$$

Simplify as far as possible:

1. $\frac{6}{9}$
2. $\frac{4}{8}$
3. $\frac{3}{15}$
4. $\frac{4}{12}$
5. $\frac{15}{30}$
6. $\frac{6}{18}$
7. $\frac{2}{4}$
8. $\frac{15}{20}$
9. $\frac{8}{12}$
10. $\frac{6}{8}$
11. $\frac{5}{20}$
12. $\frac{9}{12}$
13. $\frac{8}{10}$
14. $\frac{9}{15}$
15. $\frac{10}{18}$
16. $\frac{14}{21}$

ADDING AND SUBTRACTING FRACTIONS

We cannot add apples and oranges unless we reclassify them both as, say, fruit.

In much the same way we cannot add halves to thirds unless we express them both as equivalent fractions with the same, or common, denominator.

For example, to find $\frac{1}{2} + \frac{1}{3}$, we can express $\frac{1}{2}$ as $\frac{3}{6}$

and $\frac{1}{3}$ as $\frac{2}{6}$

so that
$$\frac{1}{2} + \frac{1}{3} = \frac{3}{6} + \frac{2}{6}$$
$$= \frac{5}{6}$$

In this example we chose 6 as the denominator for the equivalent fractions because 6 is a common multiple of the original denominators, 2 and 3. We could have chosen 12 or any other common multiple of 2 and 3, but it is sensible to keep the numbers involved as low as possible.

To subtract fractions also, we have to write them as equivalent fractions with a common denominator.

EXERCISE 2d

Find $\frac{1}{3} + \frac{3}{8}$

$$\frac{1}{3} + \frac{3}{8} = \frac{8}{24} + \frac{9}{24}$$
$$= \frac{17}{24}$$

Find:

1. $\frac{1}{3} + \frac{2}{5}$
2. $\frac{2}{5} + \frac{1}{8}$
3. $\frac{1}{3} + \frac{1}{4}$
4. $\frac{1}{2} + \frac{1}{3}$
5. $\frac{2}{3} + \frac{1}{5}$
6. $\frac{3}{4} + \frac{1}{8}$
7. $\frac{2}{3} + \frac{1}{9}$
8. $\frac{1}{6} + \frac{3}{8}$
9. $\frac{2}{5} + \frac{3}{10}$

Find $\frac{2}{9} - \frac{1}{12}$

$$\frac{2}{9} - \frac{1}{12} = \frac{8}{36} - \frac{3}{36}$$
$$= \frac{5}{36}$$

Find:

10. $\frac{2}{3} - \frac{1}{2}$
11. $\frac{3}{5} - \frac{1}{4}$
12. $\frac{5}{9} - \frac{1}{6}$
13. $\frac{5}{8} - \frac{1}{6}$
14. $\frac{5}{9} - \frac{5}{12}$
15. $\frac{3}{5} - \frac{4}{9}$
16. $\frac{5}{7} - \frac{4}{21}$
17. $\frac{2}{3} - \frac{3}{8}$
18. $\frac{5}{6} - \frac{1}{4}$

The answer to a fraction question should always be given in the lowest possible terms. Look at your answers and simplify them whenever possible.

19. $\frac{1}{3} - \frac{1}{6}$
20. $\frac{2}{5} + \frac{1}{10}$
21. $\frac{3}{8} + \frac{1}{6}$
22. $\frac{3}{4} + \frac{1}{12}$
23. $\frac{11}{12} - \frac{1}{6}$
24. $\frac{7}{12} + \frac{1}{6}$
25. $\frac{2}{3} - \frac{2}{9}$
26. $\frac{3}{8} + \frac{1}{4}$
27. $\frac{3}{5} - \frac{1}{10}$

> Find $\frac{3}{4} - \frac{1}{3} + \frac{1}{12}$
>
> $$\frac{3}{4} - \frac{1}{3} + \frac{1}{12} = \frac{9}{12} - \frac{4}{12} + \frac{1}{12}$$
> $$= \frac{\cancel{6}^{\;1}}{\cancel{12}_{\;2}}$$
> $$= \frac{1}{2}$$

Find:

28. $\frac{1}{2} - \frac{1}{3} + \frac{3}{4}$

29. $\frac{1}{6} - \frac{1}{5} + \frac{2}{3}$

30. $\frac{1}{4} + \frac{1}{2} - \frac{3}{8}$

31. $\frac{3}{5} + \frac{1}{25} - \frac{23}{50}$

32. $\frac{2}{9} - \frac{1}{3} + \frac{5}{6}$

33. $\frac{1}{4} + \frac{5}{6} - \frac{2}{3}$

34. $\frac{5}{6} - \frac{5}{18} + \frac{1}{9}$

35. $\frac{5}{7} - \frac{5}{14} + \frac{1}{2}$

36. $\frac{2}{3} + \frac{1}{2} - \frac{5}{9}$

37. $\frac{3}{10} + \frac{3}{100} - \frac{3}{50}$

38. $\frac{1}{9} + \frac{5}{6} - \frac{2}{3}$

39. $\frac{5}{8} - \frac{3}{5} + \frac{3}{10}$

MIXED NUMBERS AND IMPROPER FRACTIONS

A mixed number is the sum of a whole number and a fraction. For example "one and a half" means $1 + \frac{1}{2}$ and is written as $1\frac{1}{2}$.

Now $1 = \frac{2}{2}$ so $1\frac{1}{2} = \frac{2}{2} + \frac{1}{2} = \frac{3}{2}$

$\frac{3}{2}$ is called an improper fraction.

Whenever the numerator of a fraction is bigger than its denominator it is called an *improper fraction*.

To change an improper fraction into a mixed number, we divide the numerator by the denominator to give the number of units. The remainder is the number of fractional parts.

For example $\qquad \frac{17}{5} = 3\frac{2}{5}$

To change a mixed number into an improper fraction we multiply the units by the denominator and then add the result to the numerator.

For example $\qquad 4\frac{1}{3} = \frac{12}{3} + \frac{1}{3} = \frac{13}{3}$

Fractions and Decimals 15

EXERCISE 2e Change the following improper fractions to mixed numbers.

1. $\frac{5}{2}$
2. $\frac{4}{3}$
3. $\frac{7}{3}$
4. $\frac{7}{4}$
5. $\frac{9}{5}$
6. $\frac{15}{7}$
7. $\frac{5}{3}$
8. $\frac{7}{2}$
9. $\frac{9}{4}$
10. $\frac{11}{5}$
11. $\frac{9}{8}$
12. $\frac{17}{3}$

Change the following mixed numbers to improper fractions.

13. $1\frac{2}{3}$
14. $2\frac{1}{2}$
15. $1\frac{3}{4}$
16. $1\frac{2}{5}$
17. $3\frac{1}{3}$
18. $2\frac{1}{4}$
19. $1\frac{1}{3}$
20. $2\frac{1}{5}$
21. $1\frac{1}{4}$
22. $3\frac{2}{7}$
23. $4\frac{1}{4}$
24. $3\frac{3}{4}$

MORE ADDITION AND SUBTRACTION

Change any mixed numbers into improper fractions before adding or subtracting. If the answer is an improper fraction it is usually converted to a mixed number.

EXERCISE 2f

Find $1\frac{3}{4} - \frac{2}{3}$

$$1\frac{3}{4} - \frac{2}{3} = \frac{7}{4} - \frac{2}{3}$$
$$= \frac{21}{12} - \frac{8}{12}$$
$$= \frac{13}{12}$$
$$= 1\frac{1}{12}$$

Find:

1. $1\frac{1}{2} + \frac{1}{4}$
2. $1\frac{1}{2} - \frac{1}{4}$
3. $1\frac{1}{3} - \frac{5}{6}$
4. $2\frac{1}{2} + \frac{2}{5}$
5. $\frac{4}{7} + 1\frac{1}{2}$
6. $2\frac{1}{3} - \frac{5}{9}$
7. $1\frac{1}{3} + \frac{3}{4}$
8. $1\frac{1}{3} - \frac{3}{4}$
9. $\frac{3}{5} + 2\frac{1}{2}$

Find:

10. $3 + 1\frac{1}{2}$

11. $2 - 1\frac{1}{2}$

12. $5 - 2\frac{2}{3}$

13. $1\frac{1}{2} - \frac{2}{3} + \frac{1}{4}$

14. $\frac{2}{3} + \frac{5}{6} - 1\frac{1}{3}$

15. $1\frac{1}{4} - \frac{2}{3} + \frac{5}{6}$

16. $4 - 2\frac{1}{3}$

17. $2\frac{1}{2} - \frac{2}{3} + 1\frac{1}{6}$

18. $3\frac{3}{4} - 1$

MULTIPLYING FRACTIONS

To multiply fractions, look for common factors in numerators and denominators and cancel them.

Next multiply the numerators together,

then multiply the denominators together.

EXERCISE 2g

Find $\frac{2}{3} \times \frac{9}{10}$

$$\frac{{}^1\cancel{2}}{\cancel{3}_1} \times \frac{\cancel{9}^3}{\cancel{10}_5} = \frac{1 \times 3}{1 \times 5}$$

$$= \frac{3}{5}$$

Find:

1. $\frac{3}{4} \times \frac{2}{5}$

2. $\frac{4}{7} \times \frac{1}{8}$

3. $\frac{2}{3} \times \frac{2}{5}$

4. $\frac{4}{5} \times \frac{2}{3}$

5. $\frac{9}{10} \times \frac{7}{12}$

6. $\frac{6}{7} \times \frac{5}{9}$

7. $\frac{6}{11} \times \frac{7}{8}$

8. $\frac{6}{25} \times \frac{10}{21}$

9. $\frac{4}{7} \times \frac{5}{12}$

10. $\frac{2}{3} \times \frac{9}{10}$

11. $\frac{8}{9} \times \frac{15}{22}$

12. $\frac{9}{10} \times \frac{5}{12}$

Remember first to change mixed numbers to improper fractions.

13. $1\frac{1}{2} \times \frac{3}{4}$

14. $2\frac{1}{3} \times \frac{4}{5}$

15. $\frac{2}{7} \times 1\frac{1}{3}$

16. $2\frac{1}{2} \times \frac{3}{10}$

17. $1\frac{7}{9} \times 1\frac{7}{8}$

18. $2\frac{2}{5} \times 3\frac{3}{4}$

19. $2\frac{2}{7} \times 2\frac{1}{10}$

20. $\frac{2}{3} \times 2\frac{1}{2}$

21. $2\frac{5}{8} \times \frac{6}{7}$

22. $4\frac{2}{5} \times \frac{6}{11}$

23. $\frac{3}{4} \times 1\frac{1}{6}$

24. $2\frac{1}{5} \times \frac{5}{7}$

DIVIDING BY A FRACTION

To divide by a fraction, change $\div$ into $\times$ and turn upside down the fraction immediately following it.

Remember that whole numbers can be written as fractions by placing them over 1,

e.g. $5 = \frac{5}{1}$ so $2\frac{1}{2} \div 5 = \frac{5}{2} \div \frac{5}{1} = \frac{\cancel{5}^1}{2} \times \frac{1}{\cancel{5}_1} = \frac{1}{2}$

EXERCISE 2h

Find $1\frac{1}{2} \div \frac{1}{6}$

$$1\frac{1}{2} \div \frac{1}{6} = \frac{3}{2} \div \frac{1}{6}$$
$$= \frac{3}{\cancel{2}_1} \times \frac{\cancel{6}^3}{1}$$
$$= \frac{9}{1}$$
$$= 9$$

Remember first to change mixed numbers to improper fractions. Find:

1. $\frac{3}{5} \div \frac{2}{3}$
2. $\frac{3}{4} \div \frac{2}{5}$
3. $\frac{5}{8} \div \frac{3}{4}$
4. $\frac{8}{9} \div \frac{2}{3}$

5. $1\frac{1}{3} \div \frac{1}{2}$
6. $\frac{3}{4} \div 2\frac{1}{3}$
7. $\frac{2}{3} \div 4\frac{1}{2}$
8. $1\frac{1}{2} \div 6\frac{1}{2}$

9. $\frac{5}{9} \div \frac{20}{21}$
10. $\frac{3}{4} \div \frac{2}{9}$
11. $\frac{2}{5} \div 1\frac{1}{4}$
12. $3\frac{1}{3} \div 1\frac{1}{2}$

13. $\frac{2}{3} \div 5$
14. $\frac{3}{4} \div 2$
15. $1\frac{1}{2} \div 6$
16. $3 \div \frac{2}{3}$

17. $1\frac{1}{2} \div 2\frac{1}{4}$
18. $6 \div 1\frac{1}{2}$
19. $3 \div 1\frac{1}{2}$
20. $2\frac{1}{2} \div 10$

21. $\frac{4}{5} \div 6$
22. $8 \div 3\frac{2}{3}$
23. $1\frac{3}{4} \div 14$
24. $3\frac{1}{2} \div 1\frac{2}{5}$

MIXED QUESTIONS

EXERCISE 2i Find:

1. $\frac{2}{3} + \frac{3}{4}$
2. $1\frac{1}{2} \times \frac{1}{4}$
3. $1\frac{2}{5} - \frac{1}{3}$
4. $\frac{5}{8} \div \frac{3}{4}$
5. $1\frac{1}{2} + \frac{2}{3}$

6. $1\frac{1}{2} - \frac{2}{3}$
7. $1\frac{1}{2} \times \frac{2}{3}$
8. $1\frac{1}{2} \div \frac{2}{3}$
9. $2 - \frac{5}{8}$
10. $1\frac{3}{4} \div 4$

11. $2 \times 1\frac{1}{5}$
12. $5\frac{1}{2} \div 1\frac{4}{7}$
13. $2\frac{1}{2} + 1\frac{3}{4}$
14. $3\frac{1}{4} + 2$
15. $2 - 1\frac{1}{4}$

16. $2 \div 1\frac{1}{4}$
17. $2\frac{1}{2} + \frac{2}{5}$
18. $2\frac{1}{2} - \frac{2}{5}$

19. $2\frac{1}{2} \times \frac{2}{5}$
20. $2\frac{1}{2} \div \frac{2}{5}$
21. $1\frac{3}{4} \times 1\frac{3}{5}$

22. $2\frac{1}{7} \div 3\frac{1}{3}$
23. $5\frac{1}{2} + 2\frac{3}{8}$
24. $2\frac{2}{9} - 1\frac{1}{12}$

In the following questions you are given several alternative answers. Write down the letter that corresponds to the correct answer.

25. The value of $\frac{1}{3} + \frac{1}{4}$ is

    A $\frac{2}{7}$ **B** $\frac{1}{7}$ **C** $\frac{1}{12}$ **D** $\frac{7}{12}$

26. The value of $\frac{3}{5} \times \frac{4}{5}$ is

    A $\frac{12}{1}$ **B** $\frac{12}{25}$ **C** $\frac{7}{10}$ **D** $\frac{7}{5}$

27. The value of $2 \div \frac{1}{2}$ is

    A 4 **B** 1 **C** $2\frac{1}{2}$ **D** $\frac{1}{4}$

28. The value of $1\frac{1}{2} - \frac{3}{4}$ is

    A $\frac{1}{4}$ **B** $\frac{3}{4}$ **C** $1\frac{3}{4}$ **D** $-\frac{1}{4}$

29. The value of $2\frac{1}{2} \times \frac{1}{5}$ is

    A $12\frac{1}{2}$ **B** $2\frac{7}{10}$ **C** $\frac{1}{2}$ **D** $\frac{6}{7}$

30. The value of $1\frac{1}{2} \div \frac{1}{3}$ is

    A $\frac{1}{2}$ **B** 3 **C** $\frac{2}{9}$ **D** $4\frac{1}{2}$

MIXED OPERATIONS

When a question has a mixture of operations remember that

a) the sign in front of a number tells you what to do with that particular number, e.g.

$\frac{1}{2} \div \frac{1}{3} \times \frac{3}{4}$ means $\frac{1}{2}$ divided by $\frac{1}{3}$ and then multiplied by $\frac{3}{4}$

so deal with the division sign first. Change it to multiplication by inverting only the fraction immediately after it,

b) multiplication or division is always done before addition and subtraction,

c) it helps if brackets are put round the fractions that have to be dealt with first.

EXERCISE 2j

Find $\frac{1}{2} + \frac{2}{3} \div 1\frac{1}{2}$

$$\begin{aligned}\frac{1}{2} + \frac{2}{3} \div 1\frac{1}{2} &= \frac{1}{2} + (\frac{2}{3} \div \frac{3}{2}) \\ &= \frac{1}{2} + (\frac{2}{3} \times \frac{2}{3}) \\ &= \frac{1}{2} + \frac{4}{9} \\ &= \frac{9}{18} + \frac{8}{18} \\ &= \frac{17}{18}\end{aligned}$$

Find:

1. $\frac{3}{4} \times \frac{1}{2} - \frac{1}{8}$
2. $\frac{5}{6} + \frac{2}{3} \times \frac{1}{5}$
3. $\frac{2}{7} \div \frac{2}{5} + \frac{3}{14}$
4. $\frac{8}{15} + \frac{2}{3} \times \frac{9}{10}$
5. $\frac{5}{7} - \frac{1}{2} \div \frac{7}{8}$
6. $\frac{3}{5} \div \frac{2}{9} \times \frac{5}{13}$
7. $\frac{2}{3} + \frac{1}{2} \times \frac{4}{5}$
8. $\frac{5}{9} \div \frac{2}{3} - \frac{1}{2}$
9. $\frac{3}{4} - \frac{1}{4} \times \frac{6}{7}$
10. $1\frac{1}{3} \times \frac{2}{3} + \frac{1}{18}$
11. $2\frac{1}{2} - \frac{1}{4} \times \frac{10}{11}$
12. $\frac{2}{3} \times 1\frac{1}{5} - \frac{2}{5}$
13. $2\frac{1}{2} + 1\frac{1}{4} \times \frac{1}{5}$
14. $1\frac{2}{3} - \frac{3}{4} - 1\frac{1}{2}$
15. $\frac{2}{5} \div 4 + 1\frac{3}{10}$
16. $2\frac{1}{4} - 1\frac{1}{2} \times \frac{1}{3}$
17. $3\frac{1}{2} \times 1\frac{1}{2} - 1\frac{1}{3}$
18. $1\frac{1}{8} \times 4 - 2\frac{3}{4}$

RECIPROCALS

When we turn a fraction upside down we are finding its *reciprocal*.

For example, the reciprocal of 3, i.e. $\frac{3}{1}$, is $\frac{1}{3}$

the reciprocal of $1\frac{1}{4}$, i.e. $\frac{5}{4}$, is $\frac{4}{5}$

EXERCISE 2k Write down the reciprocals of the following numbers.

1. 5
2. 8
3. $\frac{2}{9}$
4. $\frac{7}{8}$
5. $1\frac{3}{4}$
6. $2\frac{1}{5}$
7. 2
8. 6
9. $\frac{8}{9}$
10. $\frac{1}{7}$
11. $1\frac{2}{5}$
12. $3\frac{1}{2}$

13. a) Write down the reciprocal of $\frac{2}{5}$

b) Calculate $1 \div \frac{2}{5}$

What do you notice about your answers to (a) and (b)?

14. a) Write down the reciprocal of $\frac{5}{6}$

b) Calculate $1 \div \frac{5}{6}$

What do you notice about your answers to (a) and (b)?

15. a) Write down the reciprocal of $1\frac{2}{3}$

b) Calculate $1 \div 1\frac{2}{3}$

What do you notice about your answers to (a) and (b)?

Questions 13 to 15 illustrate a general rule:

> the reciprocal of any number a, is $\frac{1}{a}$

and

> $1 \div a$ can be written as $\frac{1}{a}$

Any division can be written as a fraction.

Consider $3 \div 2$ $3 \div 2 = \frac{3}{1} \div \frac{2}{1} = \frac{3}{1} \times \frac{1}{2} = \frac{3}{2}$

i.e.

> $a \div b$ can be written as $\frac{a}{b}$

Fractions and Decimals 21

DECIMALS

We know that fractions can be written as $\frac{1}{2}$, $\frac{2}{5}$, etc. We can also represent fractions by placing a point after the unit position and continuing to add figures to the right.

The first figure after the point is the number of tenths,

e.g. $$0.5 = \frac{5}{10}$$

The second figure after the point is the number of hundredths,

e.g. $$0.57 = \frac{5}{10} + \frac{7}{100}$$ and so on.

INTERCHANGING DECIMALS AND FRACTIONS

EXERCISE 21

Express 0.12 as a fraction.

$$0.12 = \frac{1}{10} + \frac{2}{100}$$
$$= \frac{10}{100} + \frac{2}{100}$$
$$= \frac{\cancel{12}^{3}}{\cancel{100}_{25}}$$
$$= \frac{3}{25}$$

Express the following decimals as fractions.

1. 0.5
2. 0.04
3. 0.8
4. 0.15
5. 0.03
6. 0.35
7. 0.125
8. 0.75
9. 0.002
10. 0.008
11. 1.25
12. 3.75
13. 0.7
14. 0.05
15. 0.25
16. 0.005

Express $\frac{3}{4}$ as a decimal.

(Remember that $\frac{3}{4} = 3 \div 4$)

$$\frac{3}{4} = 0.75$$

$$\begin{array}{r} 0.7\;5 \\ 4\overline{)3.0^{2}0} \end{array}$$

Express the following fractions as decimals.

17. $\frac{1}{2}$ **21.** $\frac{3}{8}$ **25.** $3\frac{1}{8}$ **29.** $\frac{3}{5}$

18. $\frac{1}{4}$ **22.** $\frac{3}{100}$ **26.** $1\frac{3}{4}$ **30.** $\frac{3}{25}$

19. $\frac{2}{5}$ **23.** $\frac{7}{50}$ **27.** $2\frac{1}{2}$ **31.** $\frac{27}{100}$

20. $\frac{7}{10}$ **24.** $\frac{5}{8}$ **28.** $1\frac{1}{4}$ **32.** $2\frac{3}{4}$

Copy and complete the following table.

	Fraction	Decimal
33.	$\frac{4}{5}$	
34.		0.25
35.	$\frac{3}{4}$	
36.		0.5
37.	$1\frac{9}{10}$	
38.		0.125

ADDITION AND SUBTRACTION OF DECIMALS

Decimals are added and subtracted in the same way as whole numbers. It is sensible to write them in a column so that the decimal points are in a vertical line. This makes sure that units are added to units, tenths are added to tenths and so on.

It is also sometimes necessary to use noughts after the decimal points so that both numbers have the same number of decimal places.

For example, $3 - 0.82$ can be written as
$$\begin{array}{r} 3.00 \\ -0.82 \\ \hline \end{array}$$

EXERCISE 2m

Find $1.5 - 0.92$

$1.5 - 0.92 = 0.58$

$$\begin{array}{r} 1.50 \\ -0.92 \\ \hline 0.58 \\ \hline \end{array}$$

Fractions and Decimals

Find:

1. $1.2 + 0.7$
2. $1.2 - 0.7$
3. $3.6 + 0.8$
4. $1.3 - 0.16$
5. $0.73 - 0.002$
6. $1.7 + 0.08$
7. $2.1 - 0.8$
8. $3.5 + 1.2$
9. $1.5 - 0.47$

10. $18.04 - 12$
11. $5 + 0.17$
12. $8 - 1.25$
13. $26.56 + 1.24$
14. $17 - 1.92$
15. $25 - 10.6$
16. $24.5 - 8$
17. $6 - 5.73$
18. $3 - 0.72$

MULTIPLICATION OF DECIMALS

One way to multiply decimals is first to convert them to fractions.

For example
$$0.02 \times 1.5 = \frac{2}{100} \times \frac{15}{10}$$
(2 d.p.) (1 d.p.)
$$= \frac{30}{1000}$$
$$= 0.030$$
(3 d.p.)

From examples like this we get the following rule.

First ignore the decimal point and multiply the numbers together.

Then add together the number of decimal places in the original numbers.

This gives the number of decimal places in the answer, including any noughts at the end.

EXERCISE 2n

Find a) 1.3×0.004 b) 2.5×0.08

a) $1.3 \times 0.004 = 0.0052$
(1 d.p.) (3 d.p.) (4 d.p.)

$$\begin{array}{r} 13 \\ \times\ 4 \\ \hline 52 \end{array}$$

b) $2.5 \times 0.08 = 0.200 = 0.2$
(1 d.p.) (2 d.p.) (3 d.p.)

$$\begin{array}{r} 25 \\ \times\ 8 \\ \hline 200 \end{array}$$

Find:

1. 1.3×0.2
2. 0.5×0.5
3. 0.02×0.3
4. 1.04×0.5
5. 2.3×0.4

6. 1.9×0.02
7. 0.05×0.008
8. 1.002×0.07
9. 0.0003×0.006
10. 1.09×0.02

11. 0.8×0.5
12. 0.1×0.1
13. 0.07×0.05
14. 1.5×0.02
15. 2.06×0.003

DIVISION BY DECIMALS

To divide a decimal by a whole number you do the same as you would with whole numbers, but add noughts after the point when needed.

For example, to find $3.7 \div 2$ we have

$$\begin{array}{r} 1.8\ 5 \\ 2\overline{)3.^17^10} \end{array}$$

Division by a decimal can be converted to division by a whole number, using the fact that

> the top and bottom of a fraction can be multiplied by the same number without changing its value.

For example
$$2.3 \div 0.2 = \frac{2.3}{0.2}$$
$$= \frac{23}{2} \qquad \left(\frac{2.3 \times 10}{0.2 \times 10}\right)$$
$$= 11.5 \qquad \begin{array}{r} 11.\ 5 \\ 2\overline{)23.^10} \end{array}$$

EXERCISE 2p Find:

1. $3.5 \div 7$
2. $3.5 \div 0.7$
3. $1.3 \div 2$
4. $1.3 \div 0.02$
5. $2.6 \div 0.4$

6. $0.7 \div 0.2$
7. $0.08 \div 0.4$
8. $1.02 \div 0.2$
9. $1.5 \div 0.05$
10. $3.06 \div 0.4$

11. $2.8 \div 1.4$
12. $3.7 \div 0.02$
13. $3.9 \div 0.3$
14. $1.02 \div 0.05$
15. $2.7 \div 0.9$

MIXED QUESTIONS

EXERCISE 2q Find:

1. $1.27 + 3.6$
2. $2.5 + 3.1$
3. $0.2 + 1.6$
4. $1.8 + 0.9$

5. $3.7 + 0.12$
6. $2.5 + 0.05$
7. $1.82 + 0.8$
8. $3.02 + 1.4$

9. $4.8 - 0.7$
10. $0.2 - 0.02$
11. $5 - 4.2$
12. $3.6 - 1.2$

13. $3.6 - 0.12$
14. $0.7 - 0.7$
15. $5.05 - 3.6$
16. $5 - 3.6$

17. 1.2×0.2
18. 0.5×0.05
19. 1.1×1.1
20. 0.2×0.2

21. 1.3×0.04
22. 0.5×1.2
23. 0.4×0.4
24. 4×0.4

25. $4.2 \div 0.6$
26. $2.8 \div 7$
27. $0.36 \div 0.9$
28. $0.012 \div 0.6$

29. $5 \div 0.1$
30. $2.7 \div 0.09$
31. $5.6 \div 0.08$
32. $0.77 \div 11$

With mixed operations remember to do multiplication and division before addition and subtraction.

33. $1.7 + 0.2 \times 1.2$
34. $2.8 \div 0.7 + 0.3$
35. $3.2 + 0.04 \div 0.2$
36. $1.6 \div 0.8 \times 0.1$
37. $0.3 - 0.4 \times 0.5$

38. $1.7 + 0.3 \times 0.8$
39. $1.8 \div 0.9 \times 0.02$
40. $4.6 - 3.2 \times 0.01$
41. $0.2 + 1.8 \div 0.3$
42. $3.6 \div 0.9 - 2.1$

CORRECTING TO A GIVEN NUMBER OF DECIMAL PLACES

To correct a number to *one decimal place* we look at the figure in the *second decimal place*. If it is 5 or more we add 1 to the number in the first decimal place. If it is less than 5 we do not alter the number in the first decimal place.

For example   1.3|72 = 1.4 correct to 1 d.p.

whereas   1.3|14 = 1.3 correct to 1 d.p.

The same rule applies when correcting to any number of decimal places, i.e. always look at the next figure.

For example   1.372|8 = 1.373 correct to 3 d.p.

It is sometimes unnecessary and often impossible to give decimal answers exactly. This is particularly true in the case of measurements. We do however need to know the degree of accuracy of an answer.

For example, if a manufacturer is asked to supply washers that are about $12\frac{1}{2}$ mm in diameter, he might think that 11 mm or $13\frac{1}{2}$ mm would be good enough.

If he has to make the diameter 12.5 mm correct to 1 decimal place then he knows what is acceptable and what is not.

Give 8.0473 correct to a) 2 d.p.   b) 1 d.p.

a) 8.0473 = 8.05   to 2 d.p.

b) 8.0473 = 8.0   to 1 d.p.

(Notice that we leave the zero in the first d.p. to indicate that the number has been corrected to 1 d.p.)

EXERCISE 2r Give the following numbers correct to 1 d.p.

1. 1.35	**3.** 0.17	**5.** 0.13	**7.** 2.373
2. 2.82	**4.** 0.08	**6.** 5.02	**8.** 5.105

Give the following numbers correct to 2 d.p.

9. 0.094	**11.** 0.125	**13.** 0.027	**15.** 52.373
10. 2.037	**12.** 10.104	**14.** 0.0053	**16.** 0.058

Fractions and Decimals 27

Give the following numbers correct to 3 d.p.

17. 8.1272
18. 2.0335
19. 4.6666
20. 0.0017
21. 0.0235
22. 0.1296
23. 0.000 92
24. 0.009 53
25. 1.0278

Give the following numbers correct to 4 d.p.

26. 0.012 733
27. 2.469 912
28. 3.800 257
29. 0.000 5326
30. 1.832 051
31. 10.882 56
32. 20.305 073
33. 1.008 020 8
34. 9.999 999 9

Give the following numbers correct to the number of decimal places indicated in brackets.

35. 82.1653 (2)
36. 20.204 (1)
37. 7.707 56 (3)
38. 0.012 76 (4)
39. 1.3607 (3)
40. 15.821 (1)
41. 27.2939 (3)
42. 0.006 914 (4)
43. 0.079 88 (3)

MIXED EXERCISES

EXERCISE 2s

1. Find a) $\frac{2}{3} + \frac{3}{4}$ b) $1.7 + 0.12$

2. Find a) $\frac{2}{3} \times \frac{3}{4}$ b) $1.7 - 0.12$

3. Find a) $\frac{3}{5} \div \frac{3}{15} \times 4$ b) 1.4×0.2

4. Find a) $0.08 \div 0.2$ b) $1\frac{1}{4} \times 0.2$

5. Write down the reciprocal of a) $\frac{2}{7}$ b) 5 c) 0.5

6. Give 3.794 correct to a) 2 d.p. b) 1 d.p.

EXERCISE 2t

1. Find a) $\frac{1}{6} + \frac{3}{8}$ b) $0.14 + 1.9$

2. Find a) $\frac{1}{6} \div \frac{3}{8}$ b) $2.8 \div 0.04$

3. Find a) $1\frac{1}{2} + \frac{3}{4} \times \frac{5}{6}$ b) $1.6 - 0.7 \times 2$

4. Find a) $2\frac{1}{4} - 1\frac{5}{12}$ b) $1.7 \times 0.3 - 0.3$

5. Write down the reciprocal of a) $\frac{2}{3}$ b) $1\frac{2}{3}$ c) 0.2

6. Give 0.0736 correct to a) 3 d.p. b) 2 d.p. c) 1 d.p.

EXERCISE 2u Each question is followed by several alternative answers. Write down the letter that corresponds to the correct answer.

1. The value of $\frac{1}{2} \times \frac{2}{3}$ is

   A $\frac{1}{3}$ **B** $\frac{3}{5}$ **C** $\frac{2}{5}$ **D** $\frac{7}{6}$

2. The value of $(0.04)^2$ is

   A 0.16   **B** 0.08   **C** 0.2   **D** 0.0016

3. The reciprocal of $\frac{2}{5}$ is

   A 1   **B** $2\frac{1}{2}$ **C** $\frac{4}{25}$ **D** 0.04

4. The value of 0.009 873 correct to 2 d.p. is

   A 0.00   **B** 0.01   **C** 0.0098   **D** 0.0099

5. The value of $1\frac{1}{2} \div \frac{1}{2}$ is

   A 3   **B** 1   **C** $\frac{3}{2}$ **D** $\frac{3}{4}$

6. The value of $0.06 \div 0.3$ is

   A 2   **B** 0.2   **C** 0.02   **D** 0.002

3 SYMMETRY AND REFLECTIONS

LINE SYMMETRY

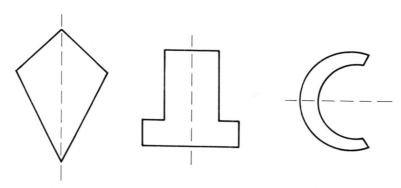

These shapes are *symmetrical*. They have line symmetry (reflective symmetry): the dotted line is the axis of symmetry, because if the shape is folded along the dotted line the two halves of the drawing fit exactly over each other.

EXERCISE 3a Copy these diagrams on to squared paper and then draw the line of symmetry. Use a dotted line or a coloured line.

1.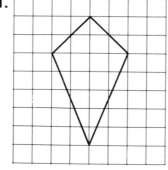

2.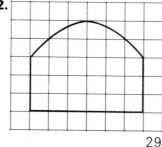

3.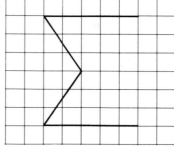

4.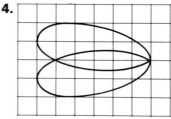

5.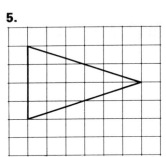

7.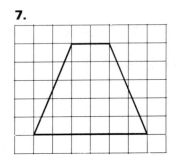

6.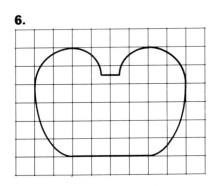

8.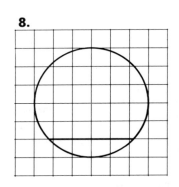

Copy the following drawings on squared paper and complete them so that the dotted line is the axis of symmetry.

9.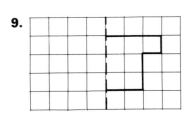

11.

10.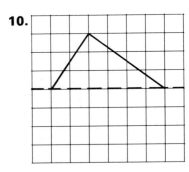

12.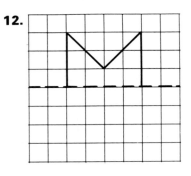

Symmetry and Reflections 31

13.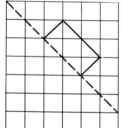

15.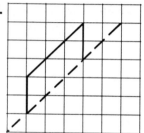

14.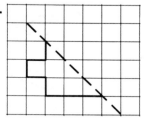

16.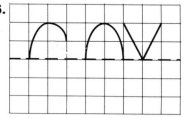

TWO OR MORE AXES OF SYMMETRY

Some shapes have more than one axis of symmetry.

A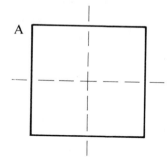

B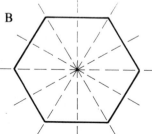

C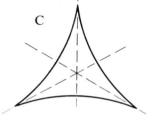

The dotted lines are the axes of symmetry. The first shape (A) has two lines of symmetry, the second shape (B) has six lines of symmetry and the third shape (C) has three lines of symmetry.

EXERCISE 3b Copy the shapes and mark in the axes of symmetry. (Some shapes may have no axis of symmetry.)

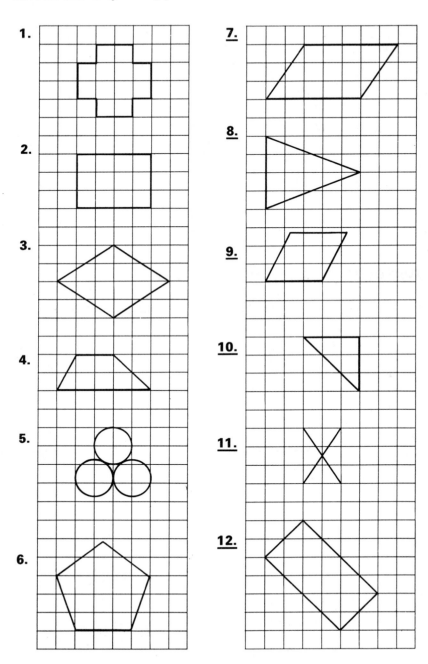

Symmetry and Reflections 33

Copy the following drawings on squared paper and complete them so that the dotted lines are the axes of symmetry.

13.
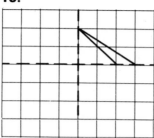

16.
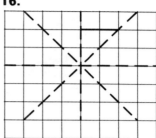

14.
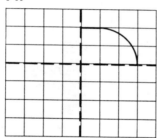

17.
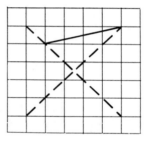

15.
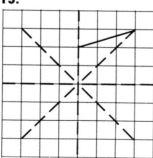

18.
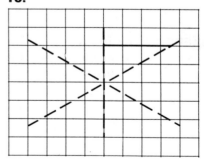

19. Draw a rectangle. How many lines of symmetry does it have?

20. Draw a square. How many lines of symmetry does it have?

REFLECTIONS

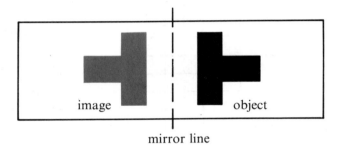

The picture shows the reflection of a drawing in a mirror.

If we imagine that the mirror is invisible we will just see the drawing and its image.

The object and its image are symmetrical, with the mirror line as the axis of symmetry.

EXERCISE 3c The dotted line is the mirror line. Copy the diagrams on to squared paper and draw the reflection of each object in the mirror line. Use a coloured pen to draw the image.

1.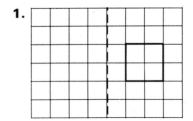

2.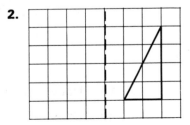

Symmetry and Reflections 35

3.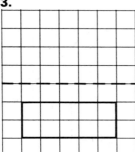

7.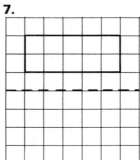

4.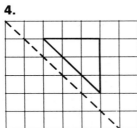

8.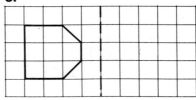

5.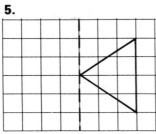

9.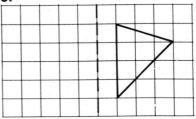

6.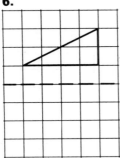

10.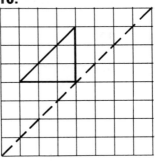

11.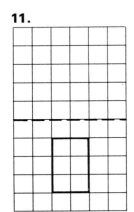

13.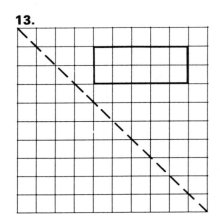

12.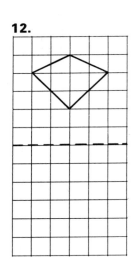

14.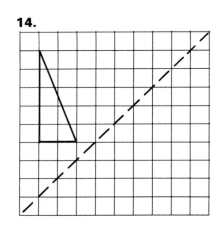

In the following diagrams the black figure is the object and the grey figure is the image. Draw in the mirror line.

15.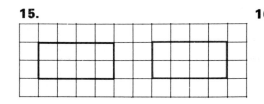

16.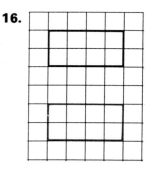

Symmetry and Reflections 37

17.
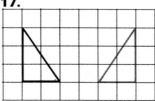

20.
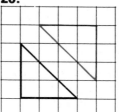

18.
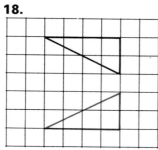

21.
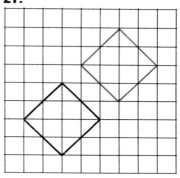

19.
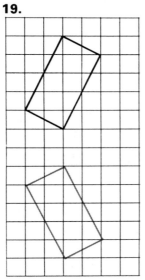

22.
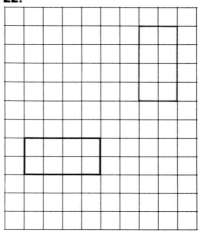

4 ROTATIONS

ROTATIONAL SYMMETRY

Some shapes have a form of symmetry that is not reflection, i.e. they do not have line symmetry but they can be rotated about a centre point and still look the same.

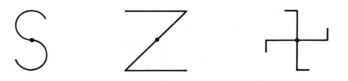

Shapes like these have rotational symmetry.

ORDER OF ROTATION SYMMETRY

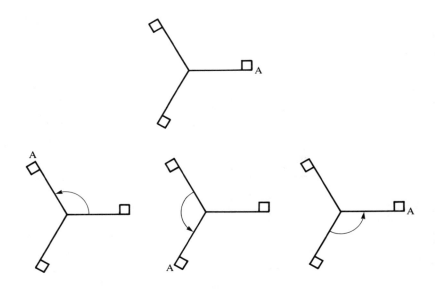

The shape in the diagram can be turned through a third of a revolution and still looks the same. If we follow one point on the shape we can see that it requires three such turns to return it to its starting position.

This shape has rotational symmetry of order 3.

EXERCISE 4a Give the order of rotational symmetry for each of these shapes.

1.

3.

5.

2.

4.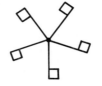

6.

ROTATIONS

We can change the position of a shape by reflecting it in a line. The reflected object is called the image.

We can also change the position of an object by rotating it about a point. The rotated object is again called the image.

In the following diagrams, the object triangle is rotated about a vertex A.

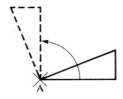

In this diagram, we get the image by rotating the object through 90° anticlockwise.

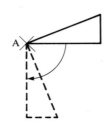

In this diagram, we get the image by rotating the object through 90° clockwise.

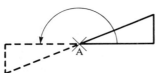

In this diagram, we get the image by rotating the object through 180° anticlockwise.

Notice that it is not enough to give just the size of the angle; the sense of turning (clockwise or anticlockwise) must also be stated.

The only case when the sense of turning is not required is for a rotation of 180°, because both a clockwise and an anticlockwise turn end with the image in the same position.

EXERCISE 4b In each of the following questions the object, line OP, is rotated about O to give the image line OP'. Give the size of the angle and the sense of turning.

1.
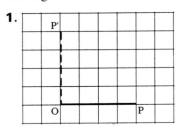

2.
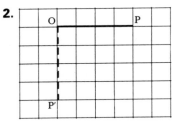

3.
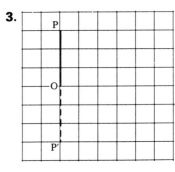

4.
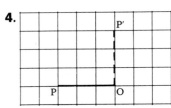

5.
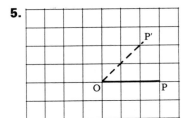

6.
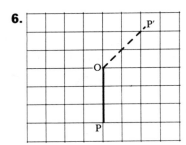

7.
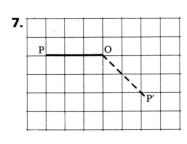

8.
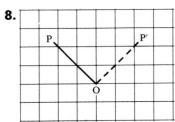

Rotations 41

In these questions, the object is the triangle ABC and it is rotated about A to give the image. Give the angle and the sense of rotation.

9.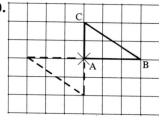

11.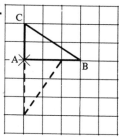

10.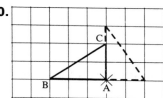

12.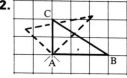

13. Copy the diagram on squared paper.

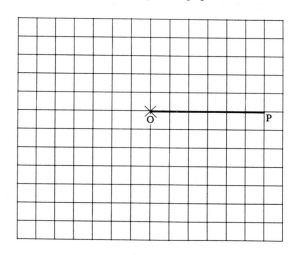

Use dotted or coloured lines to draw the three images of the line OP when it is rotated about O through

a) 90° clockwise   b) 45° anticlockwise   c) 180°

14. Copy the diagram on squared paper.

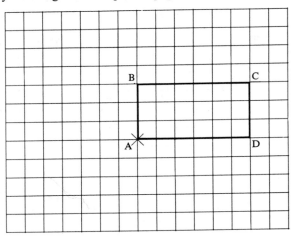

Draw the image of rectangle ABCD when it is rotated about A through 90° anticlockwise.

15. Repeat question 14 for a rotation about A of 45° clockwise.

16. Repeat question 14 for a rotation about A of 135° anticlockwise.

THE CENTRE OF ROTATION

The point about which the rotation takes place is called the centre of rotation. This point is not always on the object.

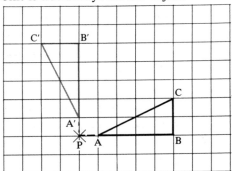

In this diagram we get the image by rotating the triangle ABC through 90° anticlockwise about P. P is the centre of rotation.

To find the angle of rotation, join a point of the object to P and join the *corresponding* point of the image to P. The angle between these two lines is the angle of rotation.

Rotations   43

EXERCISE 4c In each of the following diagrams, the image is obtained by rotating the object about P as centre. Find the angle of rotation.

1.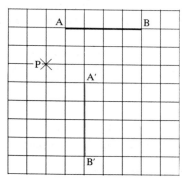

4.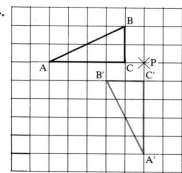

2.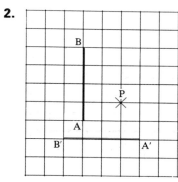

5.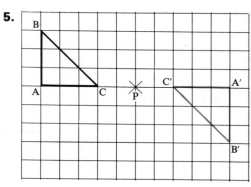

3.

6.

44   ST(P) Mathematics 3B

FINDING THE IMAGE

To find the image of an object under a given rotation, join one point of the object to the centre of rotation. Thinking of this new line as part of the object helps you to see where the image is after the required rotation.

Another way to find the image uses tracing paper and a pin. Lay the tracing paper over the object. Trace the object and, with the pin in the centre of rotation, the tracing paper can be rotated through the required angle to give the position of the image.

To find the image of rectangle ABCD under a rotation of 90° clockwise about P, we can join A to P. Thinking of AP as part of the object, it is then easy to see where the image is.

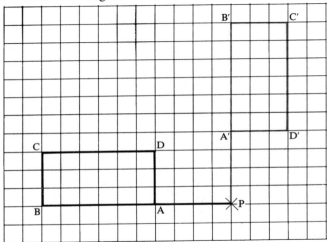

EXERCISE 4d In each of the following questions copy the diagram on squared paper. Draw the image of the given object under a rotation about P of the angle described.

1. 90° clockwise

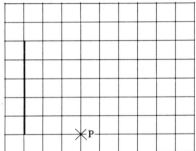

2. 90° anticlockwise

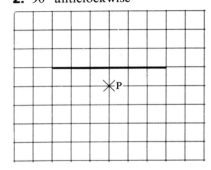

Rotations 45

3.

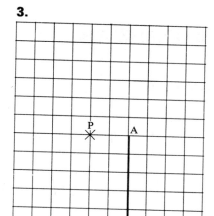

180°

5.

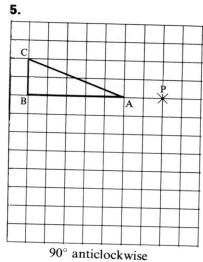

90° anticlockwise

4.

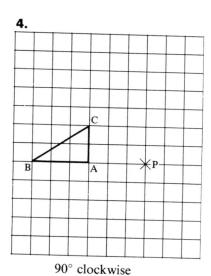

90° clockwise

6.

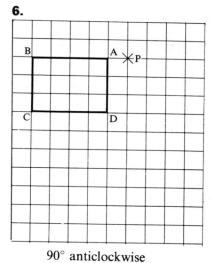

90° anticlockwise

5 DIRECTED NUMBERS

POSITIVE AND NEGATIVE NUMBERS

We use numbers to describe quantities. For example we may talk about $\frac{2}{3}$ of a cake, 10.5 cm of wire, 35 apples, and so on.
These numbers, $\frac{2}{3}$, 10.5 and 35, are examples of *positive numbers*.
We cannot however use positive numbers to describe temperatures below 0°C (the freezing point of water), or any other quantity that can fall below a zero level.
To do this we need *negative numbers*.

Positive numbers are written, for example, as $+2$ or simply as 2.
Negative numbers are written, for example, as -2.

Positive numbers and negative numbers are together known as *directed numbers*.

USING A NUMBER LINE

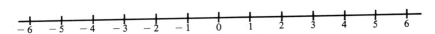

On the number line, positive numbers are in positions to the right of zero and negative numbers are in positions to the left of zero.

If a number a is to the *right* of a number b, then

a is *greater* than b

e.g. $3 > 1$ and $-1 > -3$

If a number c is to the *left* of a number d then

c is less than d

e.g. $2 < 6$ and $-6 < -4$

EXERCISE 5a Insert $>$ or $<$ between each of the following pairs of numbers.

1. 4   2
2. 3   5
3. -1 -4
4. -5 -2
5. 2 -1
6. -4 3
7. -5 -6
8. 0 -4
9. 5 -5
10. 4 -2
11. -3 0
12. 0   6

ADDING AND SUBTRACTING DIRECTED NUMBERS

Adding directed numbers can be interpreted as adding steps to the left or right,

e.g. $+(+2)$ can mean "add 2 steps to the right"

therefore $+(+2) = +2$

Also $+(-2)$ can mean "add 2 steps to the left"

therefore $+(-2) = -2$

Subtracting directed numbers can be interpreted as taking away steps to the left or right,

e.g. $-(+2)$ can mean "take away 2 steps to the right"

i.e. "go 2 steps to the left"

therefore $-(+2) = -2$

Also $-(-2)$ can mean "take away 2 steps to the left"

i.e. "go 2 steps to the right"

therefore $-(-2) = +2$

i.e. subtracting a negative number has the same effect as adding a positive number.

EXERCISE 5b

Find $+2 + (-4)$

$$+2 + (-4) = +2 - 4$$
$$= -2$$

($+2-4$ means "go 2 steps to the right and then 4 steps to the left")

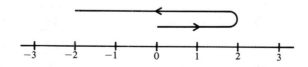

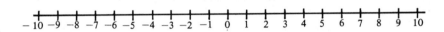

Use this number line, if it helps, to calculate:

1. $-2 + (-3)$
2. $+4 - (+6)$
3. $-2 + (-4)$
4. $-8 + (+6)$

5. $-3 - (-4)$
6. $+2 - (-6)$
7. $-5 + (-4)$
8. $3 - (-3)$

9. $-3 - (+3)$
10. $+2 - (-5)$
11. $-6 + (-4)$
12. $-2 + (+8)$

13. $-3 + (-2)$
14. $+4 - (-2)$
15. $-3 - (+2)$
16. $+2 - (+5)$

17. $+7 + (-4)$
18. $+8 - (+10)$
19. $-4 - (-8)$
20. $+6 + (-9)$

21. $-5 - (+7)$
22. $+4 + (-10)$
23. $-8 - (-3)$
24. $+10 - (+6)$

Remember that the + sign is often left out, i.e. $2 + 8 - 4$ means $+2 + (+8) - (+4)$

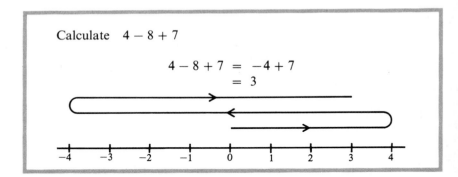

Calculate $4 - 8 + 7$

$$4 - 8 + 7 = -4 + 7$$
$$= 3$$

Calculate:

25. $3 - 4 + 6$
26. $2 + 6 - 4$
27. $-3 + 2 - 4$
28. $-4 + 2 - (-4)$
29. $5 - 3 + 9$
30. $3 - (-2) + 2$

31. $-7 + 9 - 2$
32. $3 + (-2) - 3$
33. $5 - 9 + 3$
34. $5 + 6 - 4$
35. $-3 - 7 + 9$
36. $2 - 7 - (-6)$

MULTIPLYING AND DIVIDING DIRECTED NUMBERS

$(+2) \times (+3)$ means "add 2 lots of $+3$"

therefore $\qquad (+2) \times (+3) = +(+6) = +6$

$(-2) \times (-3)$ means "take away 2 lots of -3"

therefore $\qquad (-2) \times (-3) = -(-6) = +6$

In the same way $\quad (+2) \times (-3) = +(-6) = -6$

and $\qquad (-2) \times (+3) = -(+6) = -6$

Dividing by, say, -2 is the same as multiplying by $-\frac{1}{2}$

e.g. $\qquad 12 \div (-2) = 12 \times (-\frac{1}{2}) = -6$

From this we can deduce the general rule that

> when multiplying or dividing
> like signs give a positive answer
> unlike signs give a negative answer.

EXERCISE 5c

Find a) $(-4)^2$ b) $6 \div (-2)$

a) $\qquad (-4)^2 = (-4) \times (-4)$
$\qquad \qquad \qquad = +16$

b) $\qquad 6 \div (-2) = 6 \times (-\frac{1}{2})$
$\qquad \qquad \qquad = -3$

Find:

1. $(3) \times (-2)$
2. $(-2) \times (-6)$
3. 7×8
4. $(-3) \times (-9)$
5. $(-2) \times (+4)$
6. 5×6
7. $(+3) \times (-4)$
8. $(-2) \times (-7)$
9. $(-2)^2$
10. $(-5) \times (+3)$
11. $(-3)^2$
12. $(+3) \times (-6)$

13. $(+6) \div (+3)$
14. $(+6) \div (-3)$
15. $(-6) \div (+3)$
16. $(-6) \div (-3)$
17. $(-8) \div (+4)$
18. $(+10) \div (-2)$
19. $(+9) \div (+3)$
20. $(-5) \div (-1)$
21. $4 \div 2$
22. $(-12) \div (+4)$
23. $(-3) \div 3$
24. $15 \div (-5)$

MIXED EXERCISE

EXERCISE 5d Find:

1. $2 + 6 - 5$
2. $(+3) \times (-4)$
3. $(-5)^2$
4. $2 + 7 - 9 + 2$
5. $(-3) \div (-1)$
6. $2 \times (-6)$
7. $2 - (-3)$
8. $(-6)^2$
9. $3 - (-4) + 6$
10. $(-5) \times 6$
11. $12 \div (-3)$
12. $2 - (-5) - 4$

13. $8 - 7 - 3$
14. $(-4) \times 7$
15. $(-9)^2$
16. $8 \div (-2)$
17. $5 - 7 - (-4)$
18. 4×5
19. $(-10) \div 5$
20. $7 + (-3) - 4$
21. $(-2) + 6 - 3$
22. $15 \div (-3)$
23. $7 \times (-2)$
24. $(-14) \div (-2)$

6 BRACKETS AND EQUATIONS

SIMPLIFYING ALGEBRAIC EXPRESSIONS

If an algebraic expression contains two terms, each with the same letter, then we can usually collect them.

For example $\quad x + x \quad$ can be written as $\quad 2x$

and $\quad\quad\quad\quad 2x + 3x \quad$ can be written as $\quad 5x$

because $\quad\quad 2x + 3x = x + x + x + x + x = 5x$

However, if the two terms contain different letters then, as we do not know what numbers the letters stand for, we cannot collect them.

For example $x + y$ cannot be simplified.

If two terms have the same letter *to the same index* they are the same and can be collected.

For example $\quad x^2 + 2x^2 \quad$ can be written as $\quad 3x^2$

If the two terms have the same letter *to different indices* they are *not the same* and cannot be collected.

For example $\quad x + x^2 \quad$ cannot be collected.

When we collect like terms we are simplifying.

EXERCISE 6a

Simplify where possible a) $x + y + 2x$ b) $2x + 3y$

a) $\quad\quad x + y + 2x = 3x + y$

b) $\quad\quad 2x + 3y \quad$ cannot be simplified.

Simplify where possible:

1. $a + a$
2. $2t + 4t$
3. $p + q$
4. $a + 2b + a$
5. $y + y^2$
6. $3a + 2b + a$
7. $x + 3y + 4x + 2y$
8. $a + b + c$
9. $x^2 + x + 3x$

> Simplify a) $x + 2y - 3x$ b) $p - (-2p) + 4$
>
> a) $$x + 2y - 3x = -2x + 2y$$
> $$= 2y - 2x$$
> (we usually write the positive term first)
>
> b) $$p - (-2p) + 4 = p + 2p + 4$$
> $$= 3p + 4$$

Simplify:

10. $5x - 3x$

11. $a + b - 2a$

12. $2p - (-p)$

13. $a - (-b)$

14. $3n - (-2n)$

15. $3x - 4x + 2$

16. $2t - (-4t) - 2$

17. $2b + 3 + b$

18. $x + 3x - 2y$

19. $c + 2d - (-3c)$

When terms are multiplied together we can write them in a slightly shorter form by omitting the multiplication sign.

For example $a \times b$ can be written as ab

$p \times p$ can be written as p^2 (using index notation)

$5a \times 2b$ can be written as $10ab$

($5a \times 2b = 5 \times a \times 2 \times b = 5 \times 2 \times a \times b = 10ab$)

EXERCISE 6b

> Simplify $2p \times q$
>
> $$2p \times q = 2pq$$

Simplify:

1. $x \times x$

2. $p \times q$

3. $2x \times x$

4. $2x \times 3x$

5. $4a \times b$

6. $5y \times 2z$

7. $3x \times 3x$

8. $5 \times 4x$

9. $2x \times 3$

Brackets and Equations 53

> Simplify a) $5x \times (-4x)$ b) $(-3x) \times (-2x)$
>
> a) $(5x) \times (-4x) = 5 \times (-4) \times x \times x$
> $= -20x^2$
>
> b) $(-3x) \times (-2x) = (-3) \times (-2) \times x \times x$
> $= 6x^2$

Simplify:

10. $x \times (-x)$

11. $2x \times (-3)$

12. $a \times (-b)$

13. $(-x) \times (-x)$

14. $(-2) \times (-3x)$

15. $(-2a) \times 2b$

16. $(-4) \times (-x)$

17. $2x \times (-2x)$

18. $5 \times (-2p)$

19. $(-p) \times (-3p)$

20. $(-3) \times 4x$

21. $x \times (-5y)$

22. $(-2y) \times (-3z)$

23. $x \times (-7y)$

24. $(-6p) \times 4q$

25. $(-3a) \times (-6b)$

In the following questions you are given several alternative answers. Write down the letter that corresponds to the correct answer.

26. $x + 4y + x$ can be written as

    A $6xy$ **B** $2x + 4y$ **C** $6x + y$ **D** $4x^2y$

27. $2x \times 3y$ can be written as

    A $5xy$ **B** $6x + y$ **C** $6xy$ **D** $5x + y$

28. $2x + y - 5x$ can be written as

    A $3x - y$ **B** $3x + y$ **C** $3y - 2x$ **D** $y - 3x$

29. $p \times 2p$ can be written as

    A $2p^2$ **B** $3p$ **C** $2p$ **D** $3p^2$

REMOVING BRACKETS

$5(2x + 3)$ means that both of the terms inside the brackets are multiplied by 5,

i.e. $\quad 5(2x + 3) = 5 \times 2x + 5 \times 3$
$\qquad\qquad\qquad = 10x + 15$

Also $6x(7x - 3)$ means that both $7x$ and -3 are multiplied by $6x$,

i.e. $\quad 6x(7x - 3) = 6x \times 7x + 6x \times (-3)$
$\qquad\qquad\qquad = 42x^2 - 18x$

EXERCISE 6c Remove the brackets and simplify:

1. $3(x + 2)$
2. $7(x + 4)$
3. $5(2x - 3)$
4. $4x(3x - 1)$
5. $4(x + 4)$
6. $-5(x + 3)$
7. $12x(2x - 1)$
8. $-3x(5x - 7)$
9. $7(2x - 3)$
10. $10(3x + 2)$
11. $6x(2x + 5)$
12. $-4x(3x - 7)$

Remove the brackets and simplify $2(x + 1) + 3(x + 5)$

$2(x + 1) + 3(x + 5) = 2x + 2 + 3x + 15$
$\qquad\qquad\qquad\qquad = 5x + 17$

Remove the brackets and simplify:

13. $4(x + 2) + 5(x + 1)$
14. $(x + 4) + 3(x + 2)$
15. $10(x + 3) + 5(x + 5)$
16. $6(x + 3) + 3(x + 1)$
17. $4(x + 2) + (x + 7)$
18. $3(x + 6) + 2(x + 1)$
19. $5(x + 3) + 2(x + 4)$
20. $7(x + 2) + (x + 3)$

Brackets and Equations 55

> Remove the brackets and simplify $5(x + 2) - (x + 3)$
>
> (Remember that $-(x + 3)$ means $(-1)(x + 3)$)
>
> $5(x + 2) - (x + 3) = 5x + 10 - x - 3$
> $= 4x + 7$

Remove the brackets and simplify:

21. $4(x + 1) - 3(x + 2)$ **23.** $5(x + 3) - (x + 4)$
22. $3(x + 7) - 2(x + 3)$ **24.** $3(x + 2) - 3(x + 1)$

> Remove the brackets and simplify $7(x + 2) - 5(x - 1)$
>
> $7(x + 2) - 5(x - 1) = 7x + 14 - 5x + 5$
>
> Note: $(-5) \times (-1) = 5$
>
> $= 2x + 19$

Remove the brackets and simplify:

25. $4(x + 1) - 3(x - 12)$ **27.** $5(x + 3) - 4(x - 7)$
26. $3(x - 7) - 2(x - 6)$ **28.** $10(x + 1) - 3(x - 4)$

> Remove the brackets and simplify $3(2x + 1) - 5(x + 2)$
>
> $3(2x + 1) - 5(x + 2) = 6x + 3 - 5x - 10$
> $= x - 7$

Remove the brackets and simplify:

29. $5(3x + 2) - 4(2x + 3)$ **31.** $3(7x - 2) - 5(3x - 5)$
30. $9(2x + 3) - 7(x - 6)$ **32.** $4(3x - 5) - 3(2x + 2)$

LINEAR EQUATIONS

The only general rule you need remember in solving the following types of equation is "whatever you do to one side you must do to the other."

EXERCISE 6d

Solve the equation $x + 9 = 15$

$$x + 9 = 15$$
Subtract 9 from each side $\quad x = 6$

Solve the following equations.

1. $x + 4 = 6$
2. $x + 7 = 12$
3. $x - 4 = 8$
4. $x - 7 = 9$
5. $x - 5 = 26$
6. $x + 7 = 10$
7. $x + 3 = 19$
8. $x - 7 = 3$
9. $x - 6 = 13$

Solve the equation $2x + 3 = 9$

$$2x + 3 = 9$$
Subtract 3 from each side $\quad 2x = 6$
Divide each side by 2 $\quad x = 3$

Solve equations 10 to 27.

10. $3x + 4 = 13$
11. $4x - 3 = 21$
12. $5x - 4 = 16$
13. $7x + 3 = 24$
14. $5x - 6 = 19$
15. $4x - 1 = 31$
16. $8x - 7 = 17$
17. $9x + 5 = 86$
18. $3x - 4 = 23$

The answers to the following questions are not whole numbers.

19. $3x + 7 = 23$
20. $5x - 4 = 9$
21. $4x + 11 = 21$
22. $7x + 2 = 20$
23. $3x - 4 = 1$
24. $8x - 11 = 39$
25. $4x - 9 = 12$
26. $11x + 2 = 20$
27. $6x - 3 = 13$

Brackets and Equations 57

EXERCISE 6e In this exercise the first step is to remove the brackets.

> Solve the equation $3(2x + 1) = 7$
>
> $$3(2x + 1) = 7$$
> Remove the brackets $\qquad 6x + 3 = 7$
> Subtract 3 from each side $\qquad 6x = 4$
> Divide each side by 6 $\qquad x = \frac{4}{6} = \frac{2}{3}$

Solve equations 1 to 18.

1. $3(x + 2) = 18$ **4.** $5(x - 1) = 10$ **7.** $4(x - 3) = 12$
2. $5(2x - 1) = 25$ **5.** $4(5x - 2) = 32$ **8.** $2(7x + 2) = 32$
3. $7(3x - 1) = 35$ **6.** $2(9x + 5) = 28$ **9.** $6(3x - 5) = 24$

The answers to questions 10–18 are not whole numbers.

10. $3(x + 5) = 19$ **13.** $4(5x - 2) = 20$ **16.** $7(5x + 2) = 19$
11. $5(2x - 3) = 7$ **14.** $6(3x - 5) = 1$ **17.** $10(3x - 4) = 7$
12. $2(5x - 2) = 5$ **15.** $3(2x + 5) = 11$ **18.** $4(9x - 2) = 3$

EXERCISE 6f In this exercise the first step is to remove any fractions.

> Solve $\frac{x}{4} = 4$
>
> $$\frac{x}{4} = 4$$
> Multiply each side by 4 $\qquad 4 \times \frac{x}{4} = 4 \times 4$
> $$x = 16$$

Solve the equations:

1. $\frac{x}{3} = 7$ **4.** $\frac{x}{2} = 4$ **7.** $\frac{x}{3} = 9$
2. $\frac{x}{5} = 9$ **5.** $\frac{x}{7} = 2$ **8.** $\frac{x}{8} = 2$
3. $\frac{x}{6} = -2$ **6.** $\frac{x}{3} = -4$ **9.** $\frac{x}{5} = -4$

ALGEBRAIC FACTORS

In the early part of this chapter we removed brackets and expanded expressions. Frequently we need to be able to do the reverse, i.e. to find the factors of an expression.

This is called *factorising*.

COMMON FACTORS

In the expression $8x + 12y$ we could write the first term as $4 \times 2x$ and the second term as $4 \times 3y$.

The 4 is a factor of both $8x$ and $12y$, i.e., 4 is a common factor.

We already know that $\quad 5(x + 2y) = 5 \times x \ + \ 5 \times 2y$
$$= 5x + 10y$$

In the same say $\quad 8x + 12y = 4 \times 2x \ + \ 4 \times 3y$
$$= 4(2x + 3y)$$

and $\quad 3a - 12b = 3 \times a \ - \ 3 \times 4b$
$$= 3(a - 4b)$$

EXERCISE 6g

> Factorise a) $4x - 8$ b) $5x - 5$
>
> a) $4x - 8 = 4(x - 2)$
> b) $5x - 5 = 5(x - 1)$

Factorise:

1. $3x + 3$
2. $10x + 5$
3. $7a - 14$
4. $6y - 3$
5. $9x - 3$

6. $12a - 8$
7. $4a + 12b$
8. $9y - 6$
9. $16x + 8$
10. $5x - 25y$

11. $18p - 27$
12. $2x + 10$
13. $8x - 4$
14. $21x + 14$
15. $10x - 15$

Brackets and Equations

> Factorise $x^2 - 6x$
>
> $$x^2 - 6x = x \times x - 6 \times x$$
> $$= x(x - 6)$$

Factorise:

16. $x^2 + 8x$

17. $a^2 + 12a$

18. $x^2 - 4x$

19. $4a^2 - 8$

20. $a^2 - 9a$

21. $5x^2 - x$

22. $4x^2 + x$

23. $14 + 7x$

24. $6p^2 - p$

25. $4y^2 + y$

26. $5 + 10a$

27. $9x^2 - x$

MIXED EXERCISES

EXERCISE 6h Simplify:

1. $2x + 3 - x$

2. $(-7x) \times (-4y)$

3. $p + 2q - 5p$

4. $5a + 3b$

Remove the brackets and simplify where possible:

5. $8(x + 2)$

6. $10(2x - 5)$

7. $5(3x + 2) - 4(2x + 1)$

8. $4(3x - 2) - 3(2x - 5)$

Solve the equations:

9. $4x + 1 = 17$

10. $\dfrac{x}{2} = 6$

11. $8(2x - 1) = 5$

12. $\dfrac{2x}{3} = 4$

Factorise:

13. $9x - 12$

14. $12x^2 - 5x$

15. $5a - 15b$

16. $18a - 15b$

17. $25a - 10$

18. $b^2 - 2b$

EXERCISE 6i In this exercise several alternative answers are given. Write down the letter that corresponds to the correct answer.

1. $5x + 2y - 8x$ can be written as

   A $3x + 2y$ **B** $3x - 2y$ **C** $2y - 3x$ **D** $-x + y$

2. $3x \times (-4)$ can be written as

   A $-12x$ **B** $12x$ **C** $-x$ **D** $-7x$

3. $9(x - 1) - 3(x + 1)$ simplifies to

   A $6x - 6$ **B** $12x - 6$ **C** $6x - 12$ **D** $6x + 6$

4. The solution of the equation $2x - 1 = 3$ is

   A $x = 3$ **B** $x = 1$ **C** $x = 2$ **D** $x = 0$

5. The value of x that satisfies the equation $\frac{2x}{3} = 8$ is

   A 12 **B** $1\frac{1}{3}$ **C** $5\frac{1}{3}$ **D** 48

6. $3 - 2(x - 1)$ can be simplified to

   A $x - 1$ **B** $5 - 2x$ **C** $1 - 2x$ **D** $2x + 1$

7 COORDINATES AND STRAIGHT LINE GRAPHS

COORDINATES

To fix the position of a point in a plane we use two axes, an *x*-axis (drawn across the page) and a *y*-axis (drawn up the page). The point where these axes cross is called *the origin*.

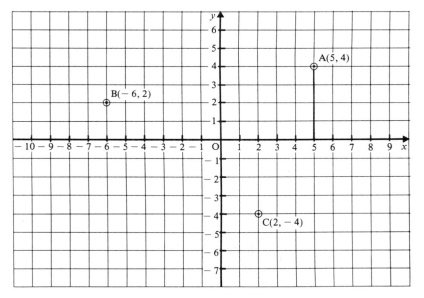

If we start from O and go 5 units to the *right*

and then go 4 units *up*

we get to the point A which we can describe by the pair of numbers (5, 4).

The numbers 5 and 4 are called *the coordinates* of A,

5 is the *x* coordinate and 4 is the *y* coordinate.

When we describe the position of a point by its coordinates, we always put the *x* coordinate first.

The point B has coordinates $(-6, 2)$. This means that, starting from O we go 6 units to the *left* and 2 units *up*.

The point C $(2, -4)$ is 2 units to the *right* and 4 units *down*.

EXERCISE 7a

1. Draw x and y axes on squared paper, numbering each axis from -6 to $+6$. Plot the points: A($-4, -3$), B($3, -3$), C($3, 1$) and D($-4, 1$).
   Join A to B, B to C, C to D and D to A.
   What is the name of the shape ABCD?

2. Draw x and y axes on squared paper, numbering each axis from -6 to $+6$. Plot the following points: A($3, -2$), B($5, -2$), C($0, 4$), D($-5, -2$), E($-3, -2$), F($-1, 0$) and G($1, 0$).
   Join the points in alphabetical order and join G to A.

3. Draw x and y axes on squared paper, numbering each one from -6 to 6. Plot the following points: A($4, 3$), B($2, 5$), C($-2, 5$), D($-4, 3$), E($-4, -3$), F($-2, -3$), G($-2, -1$), H($2, -1$), I($2, -3$), J($4, -3$).
   Join the points in order and join J to A.

LINES PARALLEL TO THE AXES

A vertical line is parallel to the y-axis and a horizontal line is parallel to the x-axis.

Consider the set of points, all of whose x coordinates are 2.

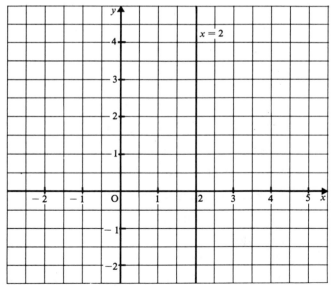

The equation of the line is $x = 2$.

This means the set of points with an x coordinate of 2.

Now consider the equation $y = 4$.

This means the set of points with a y coordinate of 4.

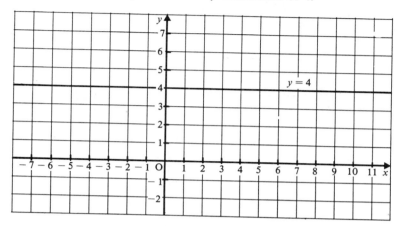

Therefore $y = 4$ represents a horizontal line which is 4 units above the origin.

EXERCISE 7b Write down the equations of the lines.

1.

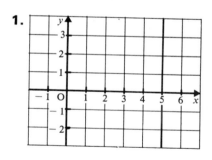

3.

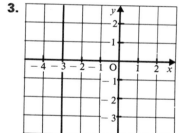

2.

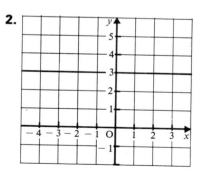

4.

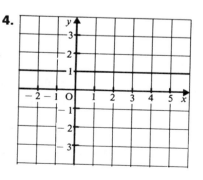

5. 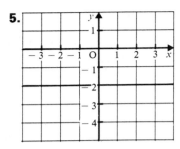   **6.**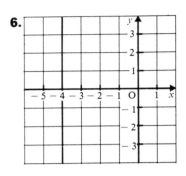

Use squared paper to draw the line with the given equation.

7. $x = 3$ **9.** $x = 2$ **11.** $x = -2$

8. $y = 4$ **10.** $y = 7$ **12.** $y = -4$

SLANT LINES THROUGH THE ORIGIN

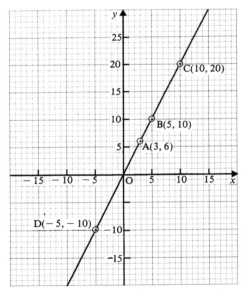

A, B, C and D are some of the points on the line in the diagram.

For each of these points

the y coordinate = 2 × the x coordinate.

Therefore the equation of the line is $y = 2x$

EXERCISE 7c

Draw x and y axes, numbering each one from -20 to $+20$ at 5-unit intervals using 1cm for 5 units on each axis.

a) Plot the points A(10, −5), B(16, −8), C(−10, 5)

b) Draw the straight line through A, B and C.

c) On your line, mark the point D whose x coordinate is 4. What is the y coordinate of D?

d) Find the connection between the x and y coordinates of each of the points A, B, C and D.

e) Write down the equation of this line.

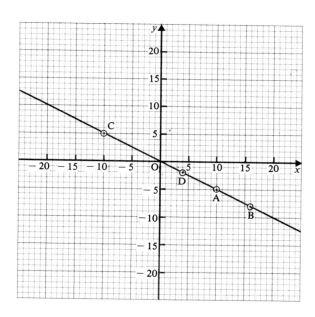

c) From the graph, the y coordinate of D is -2.

d) The y coordinate is half the size of the x coordinate, but has the opposite sign,
i.e., the y coordinate $= -\frac{1}{2} \times$ the x coordinate.

e) $y = -\frac{1}{2}x$

Use graph paper for questions 1 to 3.

Draw x and y axes, numbering each one from -20 to $+20$ at 5-unit intervals using 1 cm for 5 units on each axis.

1. a) Plot the points A(5, 5), B(18, 18), C(-10, -10).

b) Draw the straight line through A, B and C.

c) On your line mark the point D whose x coordinate is 10. What is the y coordinate of D?

d) What is the connection between the x and y coordinates for each of the points A, B, C and D?

e) Write down the equation of this line.

2. a) Plot the points A(1, 3), B(5, 15), C(-4, -12).

b) Draw the straight line through A, B and C.

c) On your line, mark the point D whose x coordinate is 2. What is the y coordinate of D?

d) What is the connection between the x and y coordinates of each of the points A, B, C and D?

e) Write down the equation of this line.

3. a) Plot the points A(10, -10), B(20, -20), C(-5, 5).

b) Draw the straight line through A, B and C.

c) On your line, mark the point D whose x coordinate is 5. What is the y coordinate of D?

d) What is the connection between the x and y coordinates for each of the points A, B, C and D?

e) Write down the equation of this line.

DRAWING A LINE FROM ITS EQUATION

Suppose that we want to draw the line whose equation is $y = 2x$.

For any point on this line,

> the y coordinate is twice the x coordinate.

Therefore if we choose a value of x, we can find the corresponding value of y.

We need only two points to draw a straight line, but it is sensible to use a third point as a check.

Coordinates and Straight Line Graphs

We will take the points whose x coordinates are -3, 0 and 2.

When $x = -3$, $y = 2 \times (-3) = -6$ so $(-3, -6)$ is on the line.
When $x = 0$, $y = 2 \times (0) = 0$ so $(0, 0)$ is on the line.
When $x = 2$, $y = 2 \times 2 = 4$ so $(2, 4)$ is on the line.

This information is usually listed in a table.

x	-3	0	2
y	-6	0	4

We now plot these points and draw the straight line through them.

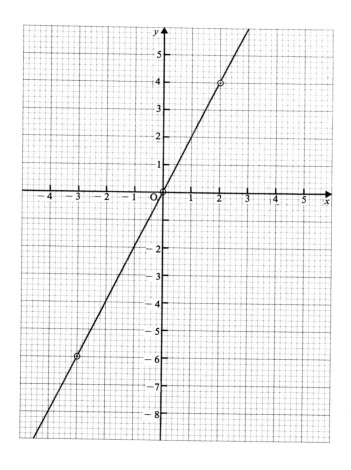

EXERCISE 7d In all questions in this exercise use graph paper and take 1cm for 1 unit on both axes.

1. The equation of a line is $y = 4x$. Copy and complete the following table to give the coordinates of three points on the line.

x	-1	0	2
y			

Draw x and y axes for the ranges $-2 \leqslant x \leqslant 3$ and $-8 \leqslant y \leqslant 9$. Plot the three points. If they are in a straight line, draw that line. If they are not in a straight line, check your arithmetic for all three points.

Repeat question 1 for the following equations, drawing x and y axes for the ranges given in brackets.

2. $y = -x$

x	0	2	5
y			

$(-2 \leqslant x \leqslant 8, \; -5 \leqslant y \leqslant 8)$

3. $y = \frac{1}{2}x$

x	-6	0	4
y			

$(-6 \leqslant x \leqslant 4, \; -3 \leqslant y \leqslant 2)$

4. $y = 6x$

x	-1	0	2
y			

$(-2 \leqslant x \leqslant 3, \; -6 \leqslant y \leqslant 12)$

5. $y = -2x$

x	-2	0	4
y			

$(-3 \leqslant x \leqslant 5, \; -8 \leqslant y \leqslant 10)$

6. $y = -4x$

x	-2	0	3
y			

$(-3 \leqslant x \leqslant 4, \; -12 \leqslant y \leqslant 10)$

7. $y = -\frac{1}{2}x$

x	-8	0	6
y			

$(-10 \leqslant x \leqslant 8, \; -4 \leqslant y \leqslant 5)$

8. $y = \frac{1}{4}x$

x	-4	0	8
y			

$(-5 \leqslant x \leqslant 9, \; -2 \leqslant y \leqslant 3)$

You are given the graph of $y = -\frac{1}{4}x$. Use the graph to find
a) y when $x = 3$
b) x when $y = 0.6$

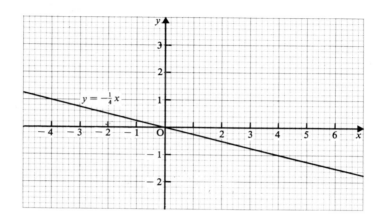

a) When $x = 3$, $y = -0.8$
b) When $y = 0.6$, $x = -2.4$

9. You are given the graph of $y = \frac{1}{2}x$. Use the graph to find
a) y when $x = 2$
b) y when $x = -2$
c) y when $x = 1.6$
d) x when $y = 2$
e) x when $y = 1.2$
f) x when $y = -0.8$

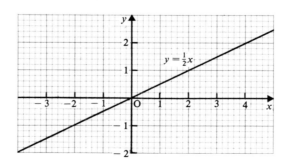

10. You are given the graph of $y = 2x$. Use the graph to find

a) y when $x = 1$

b) y when $x = -1$

c) y when $x = 1.5$

d) x when $y = 0$

e) x when $y = 1.6$

f) x when $y = -2.4$

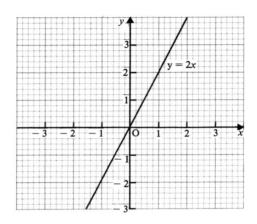

11. You are given the graph of $y = -x$. Use the graph to find

a) y when $x = 2$

b) y when $x = -1$

c) y when $x = 3.4$

d) x when $y = -3$

e) x when $y = -2.6$

f) x when $y = 1.8$

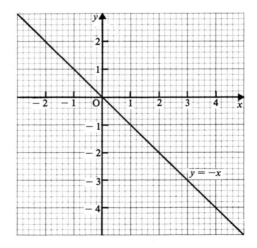

FINDING WHETHER A GIVEN POINT IS ON A LINE

Suppose that we want to find out if the point A(5, 2) is on the line with equation $y = 2x$.

The line with equation $y = 2x$ is the set of points for which the y coordinate is twice the x coordinate.

For A, the y coordinate is 2 and the x coordinate is 5.

2 is not equal to 2×5.

Therefore A is not on the line.

EXERCISE 7e

Do the points $(2, 8)$ and $(-4, 20)$ lie on the line whose equation is $y = -5x$?

For $(2, 8)$, $x = 2$ and $y = 8$
On $y = -5x$, when $x = 2$, $y = (-5) \times 2 = -10$
Therefore $(2, 8)$ is not on the line $y = -5x$

For $(-4, 20)$, $x = -4$ and $y = 20$
On $y = -5x$, when $x = -4$, $y = (-5) \times (-4) = 20$
Therefore $(-4, 20)$ is on the line.

1. Is $(2, 3)$ a point on the line whose equation is $y = x$?

2. Is $(-1, -1)$ a point on the line whose equation is $y = x$?

3. Is $(2, -2)$ a point on the line whose equation is $y = x$?

4. Is $(4, 8)$ a point on the line whose equation is $y = \frac{1}{2}x$?

5. Is $(3, 1\frac{1}{2})$ a point on the line whose equation is $y = \frac{1}{2}x$?

6. Is $(-10, -5)$ a point on the line whose equation is $y = \frac{1}{2}x$?

7. Is $(3, 6)$ a point on the line whose equation is $y = 2x$?

8. Is $(2, 2)$ a point on the line whose equation is $y = -x$?

9. Is $(2, -4)$ a point on the line whose equation is $y = -2x$?

10. Is $(8, 2)$ a point on the line whose equation is $y = \frac{1}{4}x$?

GRADIENTS

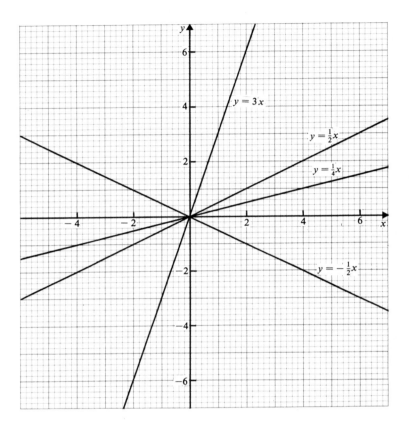

The lines in the diagram all have different slopes. We use the word *gradient* to describe slope.

If you start at a point on a line and move along the line to the right, you will find that, for the three lines with equations $y = \frac{1}{4}x$, $y = \frac{1}{2}x$ and $y = 3x$, you are moving uphill.

We say that these lines have *positive gradients*.

However, for the line with equation $y = -\frac{1}{2}x$, you will find that as you move to the right you are moving downhill.

We say that the line $y = -\frac{1}{2}x$ has a *negative gradient*.

A horizontal line has a *zero gradient*.

EXERCISE 7f

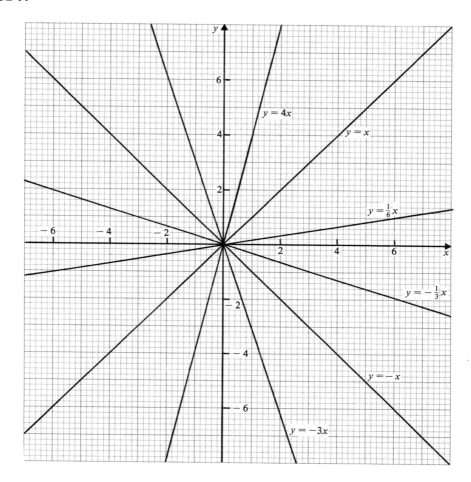

1. Use the diagram to write down
   a) the equations of the lines with positive gradients
   b) the equations of the lines with negative gradients.

2. Is there any connection between the equation of a line and the fact that it has a positive or a negative gradient?

3. State whether each of the following lines has a positive or a negative gradient. Do not draw the lines.
   a) $y = 10x$
   b) $y = \frac{1}{5}x$
   c) $y = -7x$
   d) $y = -\frac{1}{4}x$

CALCULATING THE GRADIENT

To work out the gradient of a line we first draw a diagram and mark two points, A and B, on the line. We next write on the diagram the change in *x* values and the change in *y* values as we move *right* across the page from A to B.

Then we work out the fraction

$$\frac{\text{change in } y \text{ values}}{\text{change in } x \text{ values}}$$

If we go *up* from A to B, the change in *y* values is positive.

If we go *down* from A to B, the change in *y* values is negative.

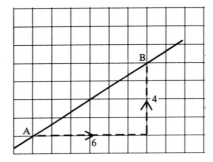

In this case the gradient is $\frac{4}{6}$ which simplifies to $\frac{2}{3}$.

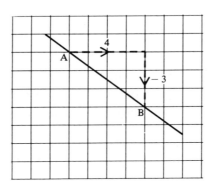

This time we have to move *down* to get from A to B, so the change in *y* values is negative.

$$\text{gradient} = \frac{-3}{4} = -\frac{3}{4}$$

EXERCISE 7g Use squared paper in this exercise and draw x and y axes for $-7 \leqslant x \leqslant 7$ and $-7 \leqslant y \leqslant 7$ using one square for 1 unit on each axis.

Find the gradient of the line joining the points
a) A(4, 2) and B(6, 7)
b) C(2, 3) and D(4, −3)

a)

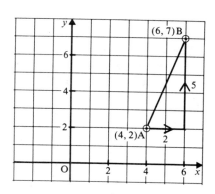

The gradient of AB is $\frac{5}{2}$

(We go *up* from A to B)

b)

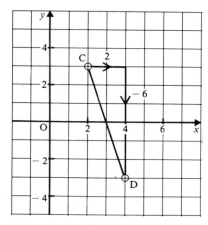

The gradient of CD is $\frac{-6}{2} = -3$

(We go *down* from C to D)

Plot both points and then find the gradient of the line joining them.

1. A(2, 3) and B(4, 5)
2. C(1, 1) and D(4, 6)
3. A(2, 2) and B(4, −5)
4. A(5, 2) and B(7, −4)
5. C(−2, 3) and D(1, 5)
6. A(1, 3) and B(4, 6)
7. C(1, 6) and D(3, 1)
8. E(2, −3) and F(5, 3)
9. A(−5, 2) and B(−1, 5)
10. M(−5, −1) and N(−1, −7)

The diagram shows the graph of the line whose equation is $y = 5x$. Points A and B are marked on the graph. Use these points to find the gradient of the line.

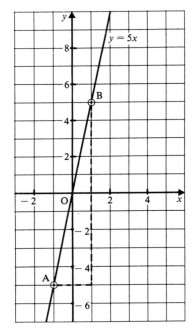

From A to B we go 2 units to the right and 10 units up.

$$\text{Gradient of AB} = \frac{10}{2}$$
$$= \frac{5}{1}$$
$$= 5$$

Coordinates and Straight Line Graphs

In each of the following questions you are given the equation of a line and its graph. Two points are marked on the line. Use these points to find the gradient of the line.

11.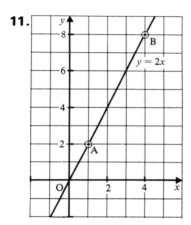

13.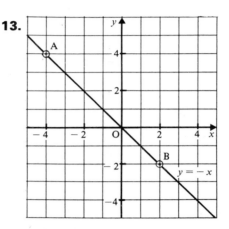

12.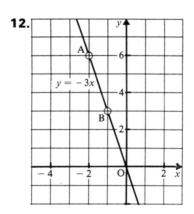

14.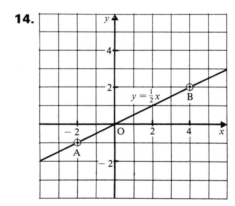

15.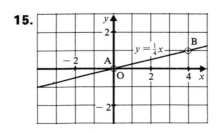

16.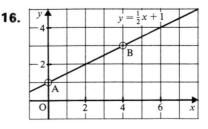

17. For questions 11 to 16 look at the value that you found for the gradient and then look at the number of xs in the equation of the line. What do you notice?

18. How does the size of the gradient affect how steep the line is?

In questions 17 to 22 you are given the equation of a line.
Do not draw the line, just write down its gradient.

19. $y = 10x$

20. $y = -\frac{7}{4}x$

21. $y = 8x$

22. $y = -\frac{1}{2}x$

23. $y = -20x$

24. $y = \frac{9}{2}x$

25. Draw the line joining the points $(4, 3)$ and $(7, 3)$.
Which axis is the line parallel to?
What is the equation of the line?
What is its gradient?

26. Plot the points $(4, 3)$ and $(4, 6)$ and draw the line joining them.
Which axis is this line parallel to?
What happens when you try to work out its gradient?

CONCLUSIONS SO FAR

From the last exercise we see that we can "read" the gradient of a line through the origin, from its equation.

For example

the line with equation $y = -\frac{7}{5}x$ has a gradient of $-\frac{7}{5}$.

In general

for a line with equation $y = mx$, the gradient is m.

We also conclude that

the larger the gradient, the steeper is the line.

If the gradient is 1, the line goes "uphill" at 45°.

If the gradient is -1, the line goes "downhill" at 45°.

EXERCISE 7h

> Sketch the line with equation $y = -\frac{3}{2}x$
>
> ("Sketch" means give a rough idea of the slope and position of the line.)
>
> (In this example $y = -\frac{3}{2}x$, therefore the gradient is negative and we see that the line slopes downhill. We see that the line slopes fairly steeply because the gradient is $-1\frac{1}{2}$.)
>
>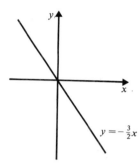

Sketch the lines whose equations are:

1. $y = x$
2. $y = -\frac{1}{2}x$
3. $y = 2x$
4. $y = -\frac{1}{4}x$
5. $y = 3x$
6. $y = -x$
7. $y = \frac{1}{2}x$
8. $y = -2x$
9. $y = \frac{1}{4}x$
10. $y = -3x$

11. $y = 4x$
12. $y = -4x$
13. $y = \frac{3}{2}x$
14. $y = -\frac{3}{2}x$
15. $y = \frac{1}{3}x$
16. $y = -\frac{1}{3}x$
17. $y = 5x$
18. $y = -5x$
19. $y = \frac{5}{2}x$
20. $y = -\frac{5}{2}x$

THE EQUATION $y = mx + c$

EXERCISE 7i

For each of the following equations, copy and complete the table to give the coordinates of three points on the line.

a) $y = x$

x	−4	0	5
y			

b) $y = x + 3$

x	−4	0	5
y			

c) $y = x − 2$

x	−4	0	5
y			

d) On the same set of axes, plot the points and draw the lines.

e) What do you notice about the three lines?

f) Where does the line with equation $y = x + 3$ cut the y-axis?

g) Where does the line with equation $y = x − 2$ cut the y-axis?

a) $y = x$

x	−4	0	5
y	−4	0	5

b) $y = x + 3$

x	−4	0	5
y	−1	3	8

$$\left\{\begin{array}{l} \text{When } x = -4, \ y = -4 + 3 = -1 \\ \text{When } x = 0, \ y = 0 + 3 = 3 \\ \text{When } x = 5, \ y = 5 + 3 = 8 \end{array}\right\}$$

c) $y = x − 2$

x	−4	0	5
y	−6	−2	3

d)

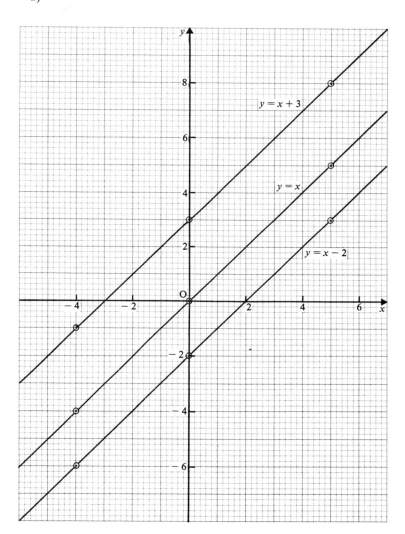

e) All three lines are parallel.

f) The line with equation $y = x + 3$ cuts the y-axis 3 units above the origin.

g) The line with equation $y = x - 2$ cuts the y-axis 2 units below the origin.

1. On graph paper draw the x and y axes for $-8 \leqslant x \leqslant 8$, $-8 \leqslant y \leqslant 8$ using 1 cm for 1 unit on each axis.

   Copy and complete the table to give the coordinates of three points on the line whose equation is given.

   a) $y = 2x$

x	-2	0	3
y			

   b) $y = 2x + 2$

x	-2	0	3
y			

   c) $y = 2x - 3$

x	-2	0	3
y			

   d) For the line given in (a), plot the points and draw the line. Using the same set of axes, do the same for the lines given in (b) and (c).

   e) What do you notice about the three lines?

   f) Where does the line whose equation is given in (a) cut the y-axis?

   g) Where does the line whose equation is given in (b) cut the y-axis?

2. Repeat question 1 for the following lines.

   a) $y = -x$

x	-4	0	4
y			

   b) $y = -x + 4$

x	-4	0	4
y			

   c) $y = -x - 1$

x	-4	0	4
y			

3.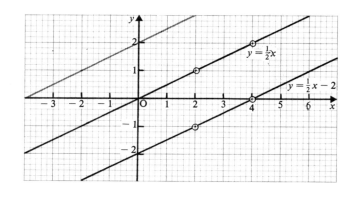

Coordinates and Straight Line Graphs   83

The diagram on the opposite page shows the graphs of the lines $y = \frac{1}{2}x$ and $y = \frac{1}{2}x - 2$.

a) Write down the gradient of the line whose equation is $y = \frac{1}{2}x$. Where does this line cut the y-axis?

b) What do you think is the gradient of the line with equation $y = \frac{1}{2}x - 2$. Use the points C and D to calculate the gradient of this line. Does your answer agree with your guess?

c) Where does the line with equation $y = \frac{1}{2}x - 2$ cut the y-axis? Is there any connection between your answer and the equation of the line?

d) Try to write down the equation of the grey line.

4.

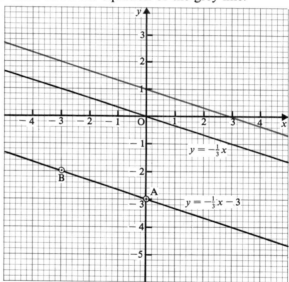

The diagram shows the graphs of the lines $y = -\frac{1}{3}x$ and $y = -\frac{1}{3}x - 3$.

a) Write down the gradient of the line $y = -\frac{1}{3}x$. Where does this line cut the y-axis?

b) Write down what you think is the gradient of $y = -\frac{1}{3}x - 3$. Use the points A and B to calculate the gradient of this line. Do your two answers agree?

c) Where does the line with equation $y = -\frac{1}{3}x - 3$ cut the y-axis? Is there a connection between your answer and the equation of the line?

d) Try to write down the equation of the grey line.

FINDING THE GRADIENT AND y INTERCEPT FROM AN EQUATION

From the last exercise we conclude that, given the equation of a line, we can "read" the gradient and the distance from O of the point where the line cuts the y-axis (called the y intercept.)

For example, if the equation of a line is $y = 5x + 3$ we know that it is parallel to the line $y = 5x$, so its gradient is 5. We also know that it cuts the y-axis 3 units above the origin.

In general

> the line with equation $y = mx + c$ has gradient m and cuts the y-axis at a point c units from the origin.
> If c is positive, the point is above the origin.
> If c is negative, the point is below the origin.

EXERCISE 7j

Sketch the line whose equation is $y = \frac{1}{3}x - 4$

$y = \frac{1}{3}x - 4$

The gradient is $\frac{1}{3}$ and the y intercept is -4

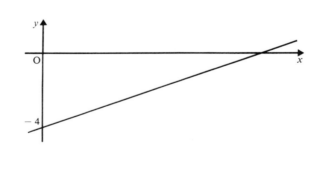

Sketch the lines whose equations are given.

1. $y = x + 2$
2. $y = -x + 3$
3. $y = 2x$
4. $y = \frac{1}{2}x + 5$
5. $y = -3x + 2$
6. $y = \frac{1}{4}x - 3$
7. $y = -x - 7$
8. $y = -2x + 4$

On the same set of axes sketch the lines whose equations are given.

9. $y = x$ **11.** $y = x - 1$ **13.** $y = x - 3$

10. $y = x + 1$ **12.** $y = x + 3$ **14.** $y = x - 5$

15. $y = -x$ **17.** $y = -x - 2$ **19.** $y = -x - 5$

16. $y = -x + 2$ **18.** $y = -x + 5$ **20.** $y = -x - 8$

FINDING THE EQUATION OF A LINE GIVEN ITS GRAPH

We know that the line with equation $y = mx + c$ has gradient m and cuts the y-axis c units from O.

This means that, given the graph of a line, we can

a) use two points on the line to work out its gradient, m

b) read where the line cuts the y-axis to find c and hence write down its equation.

EXERCISE 7k

From the given graph

a) use the points A and B to find the gradient of the line.

b) find where the line cuts the y-axis.

c) write down the equation of the line.

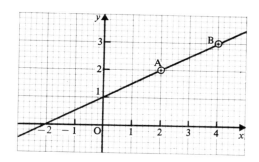

a) (From A to B we have to go 2 units right and 1 unit up.)
Gradient $= \frac{1}{2}$

b) The line cuts the y-axis 1 unit above O.

c) The equation is $y = \frac{1}{2}x + 1$

For each of the following graphs find
a) the gradient of the line (use points A and B if necessary)
b) where the line cuts the y-axis
c) the equation of the line.

1.
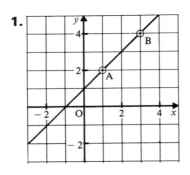

3.
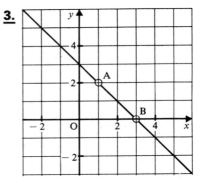

2.
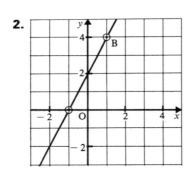

4.
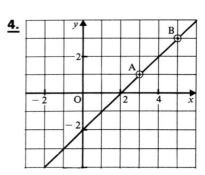

Write down the equation of each of the following lines. (If necessary, choose two points to work out the gradients.)

5.
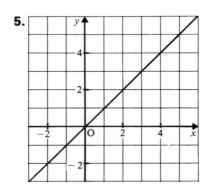

6.
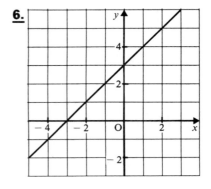

7.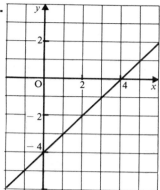

10.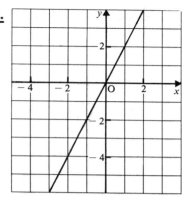

8.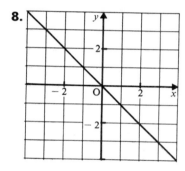

11.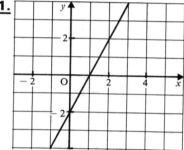

9.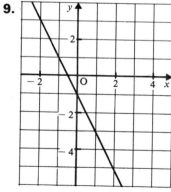

12.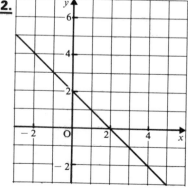

INTERSECTION OF TWO LINES

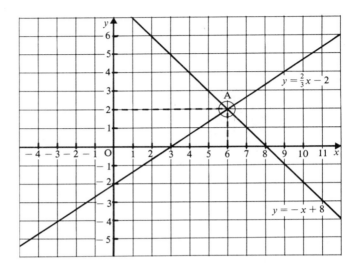

The diagram shows the lines whose equations are

$$y = \tfrac{2}{3}x - 2 \quad \text{and} \quad y = -x + 8$$

The point A where they cross is called the *point of intersection* of the two lines.

This is the only point that lies on both lines. Its coordinates, from the graph, are $(6, 2)$.

We say that $x = 6$ and $y = 2$ satisfies both of the equations $y = \tfrac{2}{3}x - 2$ and $y = -x + 8$

EXERCISE 71 In questions 1 and 2, A is the point of intersection of the two given lines. Write down the coordinates of A.

1.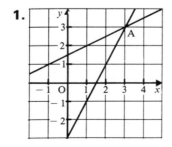

2.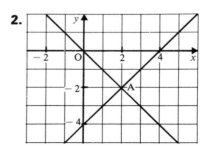

In questions 3 to 5, copy the given diagram on to squared paper.

3.

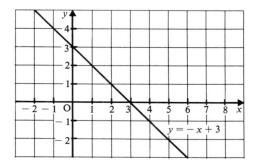

a) The equation of a second line is $y = \frac{1}{2}x$. For this line copy and complete the table

x	−2	0	6
y			

b) Using the diagram that you copied, plot these points and draw the line joining them.

c) Write down the coordinates of the point of intersection of the two lines.

4.

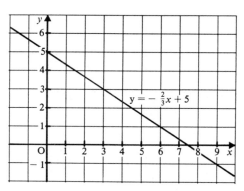

a) The equation of a second line is $y = x + 2$. For this line copy and complete the table

x	0	6	9
y			

Repeat parts (b) and (c) of question 3.

5.

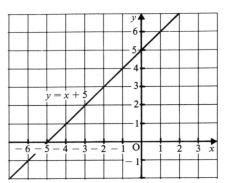

a) The equation of a second line is $y = -x + 1$. For this line, copy and complete the table

x	-5	0	2
y			

Repeat parts (b) and (c) of question 3.

MIXED EXERCISES

EXERCISE 7m **1.** Write down the equations of the following lines.

a)

b)

c)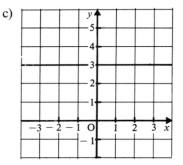

d)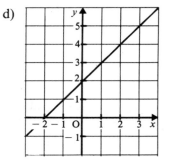

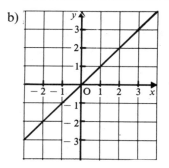

e) 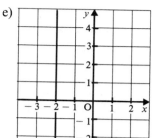 f)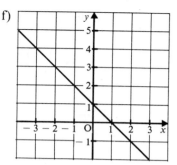

2. Use squared paper and draw axes for x and y in the ranges $-6 \leqslant x \leqslant 6$, $-6 \leqslant y \leqslant 6$, using 1cm for 1 unit on each axis. Plot each pair of points and find the gradient of the line joining them.

    a) $(1, 2)$ and $(2, 3)$ c) $(-5, 4)$ and $(2, -1)$
    b) $(1, 4)$ and $(2, 6)$ d) $(-4, -4)$ and $(-5, 2)$

3. The point $(1, 2)$ lies on two of the lines whose equations are $y = x + 1$, $y = x + 4$, $y = -x + 3$ and $y = 2x - 1$. Write down the equations of these two lines.

4.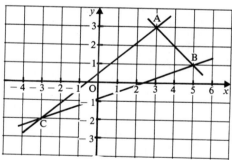

   A, B and C are the points of intersection of the graphs shown. Write down the coordinates of A, B, and C.

5. On squared paper draw x and y axes in the ranges $-4 \leqslant x \leqslant 10$, and $-4 \leqslant y \leqslant 6$, using 1cm for 1 unit on each axis. On these axes draw the graphs whose equations are

    a) $y = 5$ b) $y = x$ c) $x = -2$ d) $y = x - 4$

6. Using graph paper draw axes for $-3 \leqslant x \leqslant 9$ and $-2 \leqslant y \leqslant 6$. Take 1cm for 1 unit on both axes.

a) For the line whose equation is $y = -\frac{1}{2}x + 4$, copy and complete the table

x	-2	0	6
y			

b) Plot the points and draw the line.

c) From the graph find the coordinates of the point where the line crosses the x-axis.

Questions 7 to 10 each show the graph of a line. Each line has one of the following equations.

A $y = x + 2$ **B** $y = -x + 3$ **C** $y = 2x$ **D** $y = x - 1$

For each question write down the letter that corresponds to the equation of the line shown.

7.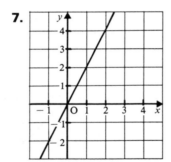

9.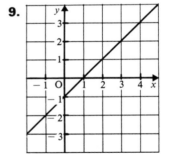

8.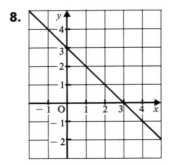

10.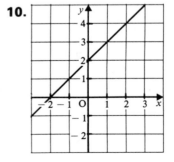

EXERCISE 7n In this exercise each question is followed by several alternative answers. Write down the letter that corresponds to the correct answer.

1.
The gradient of the line shown is

A 2   B 1
C −1   D −2

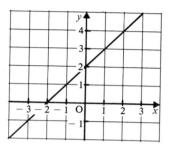

2. One of the following points lies on the line whose equation is $y = x + 4$. Which one is it?

A (0, 4)   B (0, 1)
C (1, 0)   D (4, 0)

3.
The equation of the line shown is

A $y = 1$ B $x = -1$
C $y = x + 1$ D $y = x - 1$

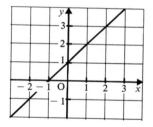

4. The point $(1, -2)$ lies on the line whose equation is

A $y = x + 2$ B $y = -x + 1$
C $y = -x - 1$ D $y = 2x + 1$

5.
The equation of the line shown is

A $y = x$ B $x = 1$
C $y = 1$ D $y = 0$

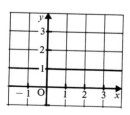

6.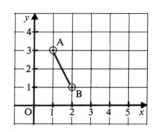

The gradient of the line AB is

A 2 **B** $\frac{1}{2}$ **C** $-\frac{1}{2}$ **D** -2

7. The equation of a line is $y = x - 3$. The graph of this line is

A 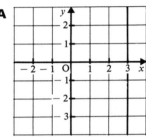   **B**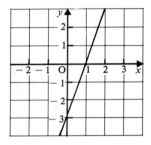

C 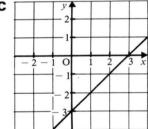   **D**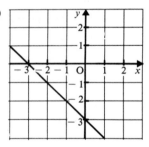

8.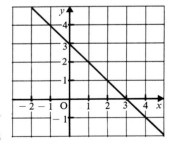

The equation of the line shown is

A $y = x + 3$ **B** $y = -x + 3$

C $y = x - 3$ **D** $y = -x - 3$

8 TRANSLATIONS

VECTORS

If you want to describe where Birmingham is in relation to London you have to give two pieces of information, the distance and the direction.

Quantities that have both size and direction are called *vectors*.

We can represent a vector by a straight line and indicate its direction with an arrow. For example

We use **a, b, c, d,**... to name the vectors.

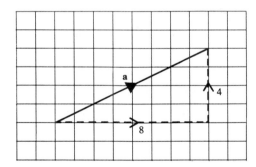

In this diagram the movement along **a** corresponds to a move of 8 units to the right and 4 units upwards, and we write

$$\mathbf{a} = \begin{pmatrix} 8 \\ 4 \end{pmatrix}$$

96   ST(P) Mathematics 3B

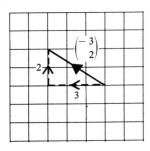

The vector $\begin{pmatrix} -3 \\ 2 \end{pmatrix}$ corresponds to a move of 3 units to the *left* and 2 units upwards.

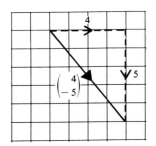

The vector $\begin{pmatrix} 4 \\ -5 \end{pmatrix}$ corresponds to a move of 4 units to the right and 5 units *downwards*.

The top number gives the horizontal move and the lower number the vertical move.

EXERCISE 8a On squared paper, draw a line to represent the following vectors.

1. $\begin{pmatrix} 3 \\ 4 \end{pmatrix}$ 5. $\begin{pmatrix} -2 \\ 4 \end{pmatrix}$ 9. $\begin{pmatrix} 7 \\ -4 \end{pmatrix}$ 13. $\begin{pmatrix} 5 \\ 4 \end{pmatrix}$

2. $\begin{pmatrix} 2 \\ 5 \end{pmatrix}$ 6. $\begin{pmatrix} -3 \\ 2 \end{pmatrix}$ 10. $\begin{pmatrix} 8 \\ 0 \end{pmatrix}$ 14. $\begin{pmatrix} -2 \\ -7 \end{pmatrix}$

3. $\begin{pmatrix} 1 \\ 6 \end{pmatrix}$ 7. $\begin{pmatrix} 0 \\ 5 \end{pmatrix}$ 11. $\begin{pmatrix} -2 \\ -4 \end{pmatrix}$ 15. $\begin{pmatrix} 0 \\ -4 \end{pmatrix}$

4. $\begin{pmatrix} 4 \\ 1 \end{pmatrix}$ 8. $\begin{pmatrix} 2 \\ -3 \end{pmatrix}$ 12. $\begin{pmatrix} 5 \\ -8 \end{pmatrix}$ 16. $\begin{pmatrix} -6 \\ 0 \end{pmatrix}$

Write down, in the form $\begin{pmatrix} x \\ y \end{pmatrix}$, the vectors which are represented by the following lines.

17.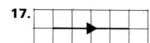

18.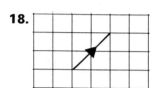

19.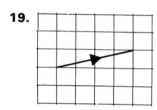

20.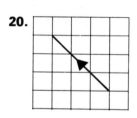

21.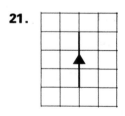

22.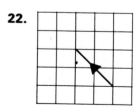

23.

24.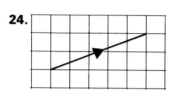

25.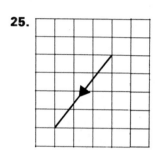

26.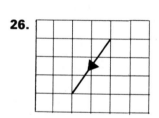

27.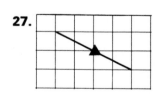

28.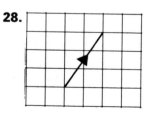

TRANSLATIONS

We know that the position of an object can be changed by reflecting it or by rotating it. Another way to change the position of an object is to slide it along a straight line. This way of producing an image is called a *translation*.

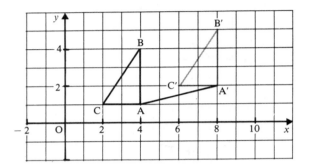

In the diagram, the image has been produced from the object, △ABC, by moving it 4 units to the right and 1 unit upwards. We can see this clearly if we look at point A of the object and the corresponding point A′ of the image.

To describe this translation we use the vector $\begin{pmatrix} 4 \\ 1 \end{pmatrix}$ as this represents a movement of 4 units to the right and 1 unit up.

EXERCISE 8b In each of the following diagrams the grey figure is the image. Join a point of the object to the corresponding point of the image and hence write down the vector which describes the translation.

1.

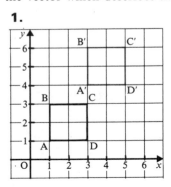

2.

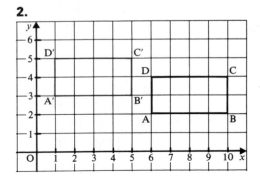

Translations 99

3.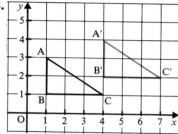

7.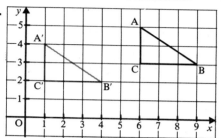

4.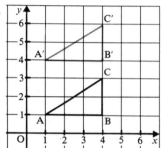

8.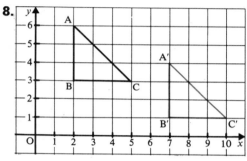

5.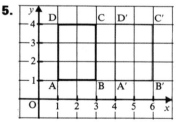

9.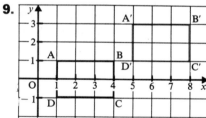

6.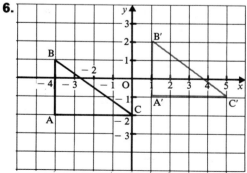

10.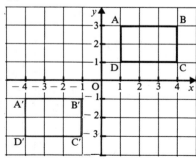

Draw the image of triangle ABC after a translation of $\begin{pmatrix} 3 \\ 4 \end{pmatrix}$

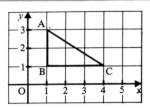

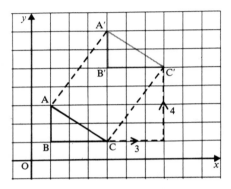

(Each point of the object is moved the same distance in the same direction, i.e. 3 squares right and 4 squares up. Move as many points as you need to, until you can see the position of the image.)

Copy the following diagrams on to squared paper. For each question use a dotted or coloured line to draw the image of the given object after a translation described by the given vector.

11.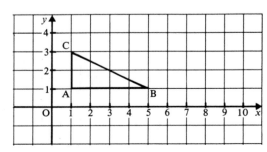

$\begin{pmatrix} 5 \\ 2 \end{pmatrix}$

12. 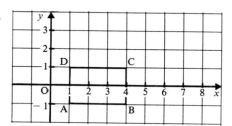 $\begin{pmatrix} 4 \\ 3 \end{pmatrix}$

13. 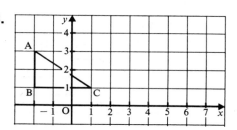 $\begin{pmatrix} 6 \\ 1 \end{pmatrix}$

14. 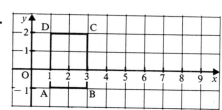 $\begin{pmatrix} 5 \\ 0 \end{pmatrix}$

15. 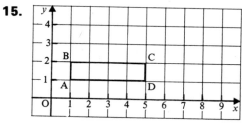 $\begin{pmatrix} 0 \\ 3 \end{pmatrix}$

16. 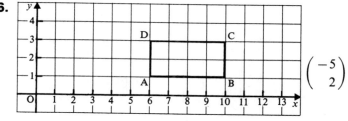 $\begin{pmatrix} -5 \\ 2 \end{pmatrix}$

17.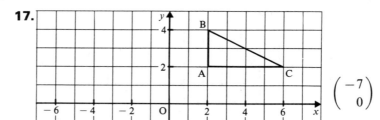

18.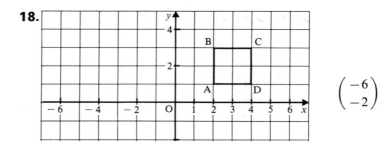

19.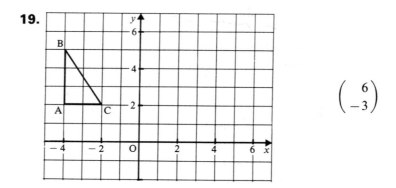

20.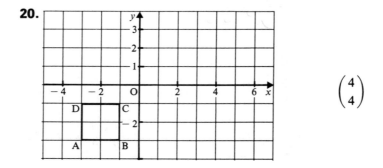

TRANSFORMATIONS

Translations 103

When an object is reflected, rotated or translated, an image is produced. We say that the object is *mapped* on to the image and the way in which it is mapped is the *transformation*.

EXERCISE 8c Copy each diagram on to squared paper. Draw the image of the given object that is produced by the given transformation.

1.

3.

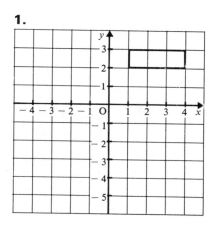
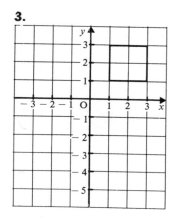

Reflection in the *x*-axis
(i.e. the *x*-axis is the mirror line)

Reflection in the *y*-axis

2.

4.

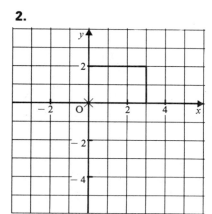
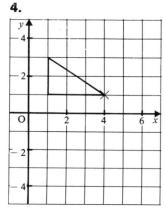

Rotation about O through
90° clockwise

Rotation about (4, 1)
through 90° anticlockwise

5.

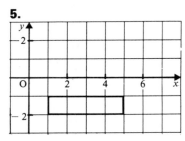

Translation of $\begin{pmatrix} 2 \\ 3 \end{pmatrix}$

8.

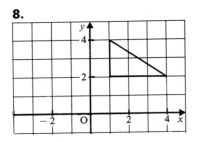

Reflection in the y-axis

6.

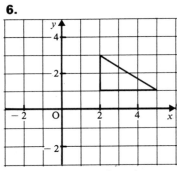

Translation of $\begin{pmatrix} -4 \\ -2 \end{pmatrix}$

9.

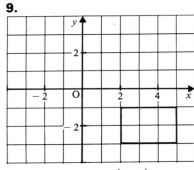

Translation of $\begin{pmatrix} -5 \\ 4 \end{pmatrix}$

7.

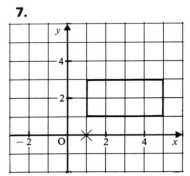

Rotation about $(1, 0)$ through $90°$ anticlockwise

10.

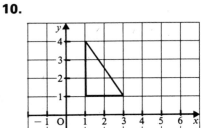

Reflection in the line $x = 3$

11.

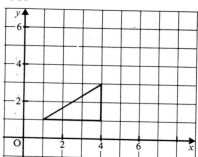

Reflection in the line $y = 3$

12.

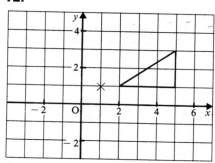

Rotation about $(1, 1)$ through $180°$

13. Copy the diagram given in question 10 extending the axes as required. On your diagram draw the three images of the triangle produced by

a) a reflection in the line $x = -1$

b) a translation of $\begin{pmatrix} 4 \\ 3 \end{pmatrix}$

c) a rotation about $(3, 0)$ through $90°$ clockwise.

14. Copy the diagram given for question 5 extending the axes as required. On your diagram draw the three images of the triangle given by

a) a reflection in the line $y = x$

b) a rotation about $(5, 0)$ through $90°$ clockwise

c) a translation of $\begin{pmatrix} -4 \\ 0 \end{pmatrix}$

9 ESSENTIAL GEOMETRY

This Chapter revises some of the basic facts of geometry.

BASIC ANGLE FACTS

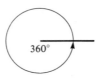 One complete revolution is 360°.

 One quarter of a revolution is 90° and is called a right angle.

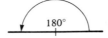 Half a revolution is 180°.

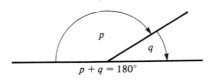

When two angles make half a revolution (i.e., make a straight line) they are called supplementary angles.

> Angles on a straight line add up to 180°

Essential Geometry 107

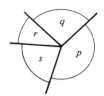

When several angles together make a complete revolution they are called angles at a point.

$p + q + r + s = 360°$

> Angles at a point add up to 360°.

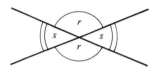

When two straight lines cross, four angles are formed. The two angles that are opposite to each other are called vertically opposite angles.

> Vertically opposite angles are equal.

EXERCISE 9a Calculate the size of the angle marked x.

1.

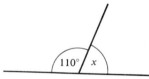

4.

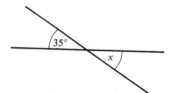

2.

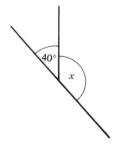

5.

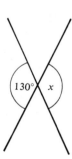

3.

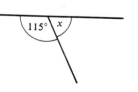

6.

7.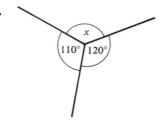

10.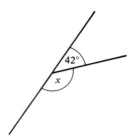

8.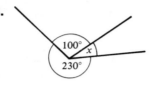

11.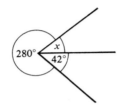

9.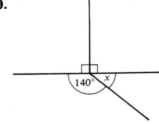

12.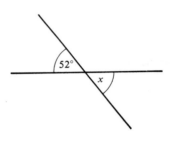

Find the size of the angle marked with a letter.

13.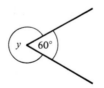

15.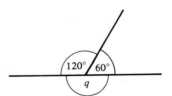

14.

16.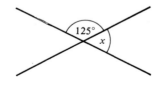

Essential Geometry   109

In the following questions you are given several alternative answers. Write down the letter that corresponds to the correct answer.

17.

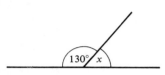

The angle marked *x* is

    A 130°   **B** 50°   **C** 230°   **D** 70°

18.

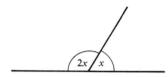

The value of *x* is

    A 60°   **B** 120°   **C** 45°   **D** 90°

19.

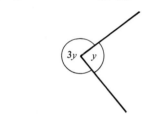

The value of *y* is

    A 180°   **B** 120°   **C** 90°   **D** 60°

20.

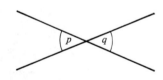

The angles *p* and *q*

    A are supplementary   **B** add up to 180°   **C** are equal

21.

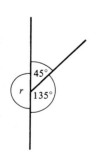

The angle marked *r* is

    A 180°   **B** 135°   **C** 45°

ANGLES AND TRIANGLES

A triangle has three angles and three sides.
The corners are called vertices and are labelled with capital letters.
Each letter stands for the vertex and also for the angle,
i.e. in the diagram we can talk about the vertex A or the angle A.

We refer to the triangle as $\triangle ABC$.

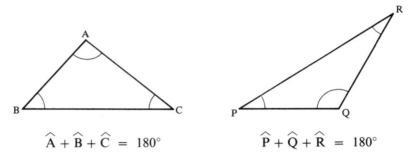

$\widehat{A} + \widehat{B} + \widehat{C} = 180°$ $\widehat{P} + \widehat{Q} + \widehat{R} = 180°$

Whatever the shape or size of the triangle

> the three angles of a triangle add up to 180°.

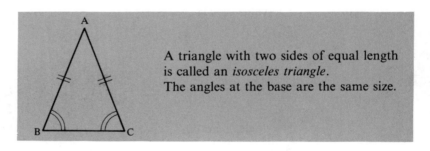

A triangle with two sides of equal length is called an *isosceles triangle*.
The angles at the base are the same size.

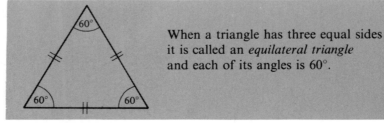

When a triangle has three equal sides it is called an *equilateral triangle* and each of its angles is 60°.

EXERCISE 9b Find the size of the angle marked x.

1.
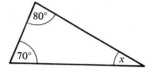

6.
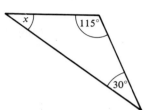

2.
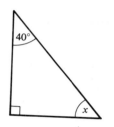

7.
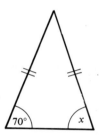

3.

8.
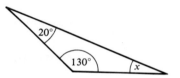

4.
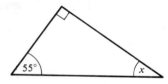

9.
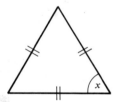

5.
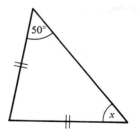

10.
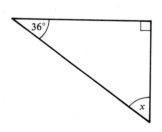

Find the size of the angle marked x.

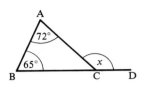
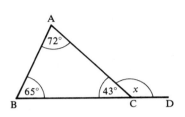

In $\triangle ABC$, $65° + 72° + \widehat{ACB} = 180°$
$137° + \widehat{ACB} = 180°$
$\widehat{ACB} = 43°$

BCD is a straight line, so $43° + x = 180°$
$x = 137°$

Find the size of the marked angle.

11.
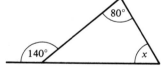

13.

12.
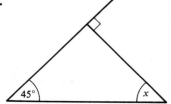

14.
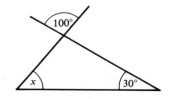

15.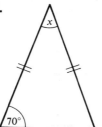

17.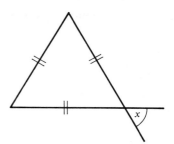

16.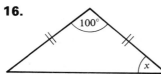

18.

19.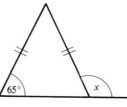

20.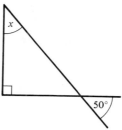

21.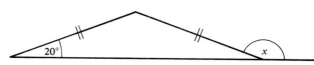

22.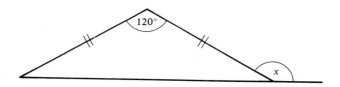

PARALLEL LINES AND ANGLES

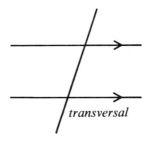

Parallel lines are always the same distance apart.

A line that cuts a set of parallel lines is called a *transversal* and it forms several angles.

These diagrams show pairs of corresponding angles.

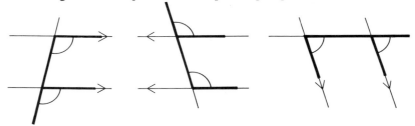

Corresponding angles are equal.

Corresponding angles can be recognised by looking for an F shape.

These diagrams show pairs of alternate angles.

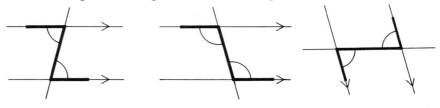

Alternate angles are equal.

Alternate angles can be recognised by looking for a Z shape.

Essential Geometry   115

These diagrams show pairs of interior angles.

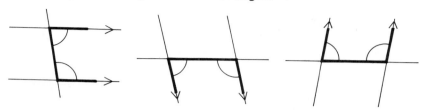

> Interior angles add up to 180°.

Interior angles can be recognised by looking for a U shape.

EXERCISE 9c   In each of the following diagrams one of the angles is shaded. Write down the letter of the angle which is

   a) corresponding   b) alternate   c) interior,

to the shaded angle.

1.
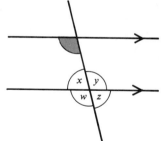

3.
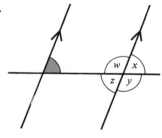

2.
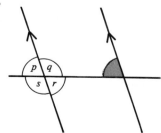

4.
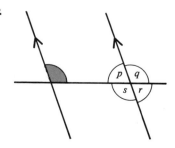

116   ST(P) Mathematics 3B

Find the size of the angle marked x.

5.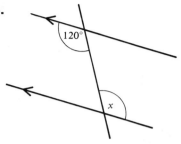

8.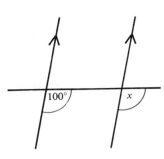

6.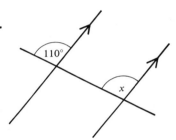

9.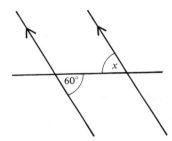

7.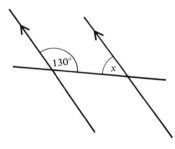

10.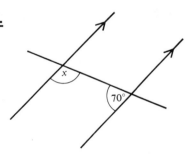

MIXED EXERCISES

EXERCISE 9d CALCULATIONS

Find the size of the marked angles. Use *any* facts that you know.

1.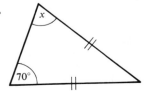

2.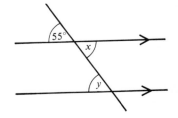

Essential Geometry   117

3.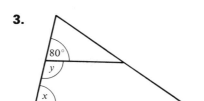

4.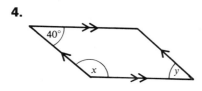

5.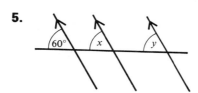

6.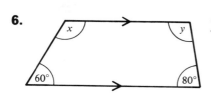

7.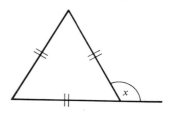

8.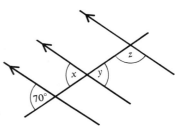

9.

10.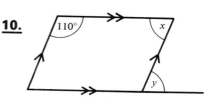

EXERCISE 9e In this exercise you have to find x. Several alternative answers are given. Write down the letter that corresponds to the correct answer.

1.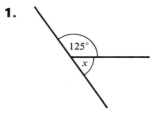

    A 125°   **B** 55°   **C** 235°

2.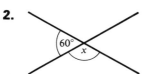

    A 120°   **B** 60°   **C** 180°

3. 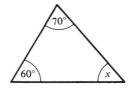   **A** 60°   **B** 50°   **C** 30°

4. 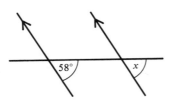   **A** 122°   **B** 29°   **C** 58°

5. 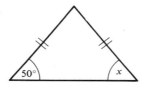   **A** 50°   **B** 130°   **C** 80°

6. 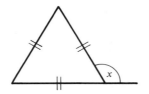   **A** 60°   **B** 30°   **C** 120°

7. 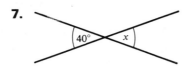   **A** 40°   **B** 140°   **C** 320°

8. 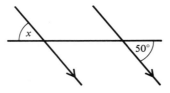   **A** 50°   **B** 130°   **C** 80°

9.    **A** 90°   **B** 110°   **C** 20°

Essential Geometry 119

EXERCISE 9f DRAWING

Use squared paper, ruler, compasses and a protractor. Note that the diagrams in this exercise are not drawn to scale.

1. Draw an equilateral triangle whose sides are 8 cm long. Measure the distance from one vertex to the opposite side. (Remember that this means the perpendicular distance.)

2.

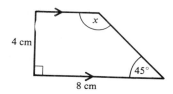

   a) Make a full size copy of the figure
   b) *Calculate* the size of angle x
   c) *Measure* angle x.

3.

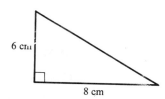

   a) Make a full size copy of the figure
   b) Measure the longest side of the triangle.

4.

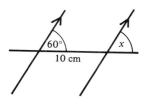

   a) Calculate the size of angle x
   b) Copy the figure, making it full size.

5.

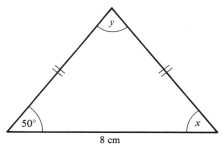

a) Write down the size of angle *x*
b) Make a full size copy of the triangle
c) *Calculate* angle *y*
d) *Measure* angle *y*.

6.

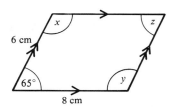

a) Write down the size of angle *x*
b) Write down the size of angle *y*
c) Make a full size copy of the parallelogram
d) Measure angle *z*.

7.

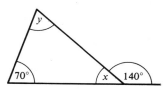

a) Calculate angle *x*
b) Make a full size copy of the figure
c) *Measure* angle *y*
d) *Calculate* angle *y*.

8.

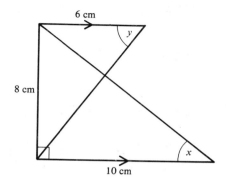

a) Make a full size copy of the figure
b) Measure angle x
c) Measure angle y.

10 PERCENTAGES

THE MEANING OF PERCENTAGE

A percentage is a fraction.

7 per cent, or 7%, means "7 per hundred", i.e. $\frac{7}{100}$

10% of a quantity is $\frac{10}{100}$, i.e. $\frac{1}{10}$, of the quantity.

CHANGING PERCENTAGES TO FRACTIONS

EXERCISE 10a

Give 4% as a fraction.

$$4\% = \frac{4}{100} = \frac{1}{25}$$

Give the following percentages as fractions. Simplify each fraction if possible.

1. 3%	**6.** 16%	**11.** 90%	**16.** 20%
2. 15%	**7.** 44%	**12.** 65%	**17.** 81%
3. 40%	**8.** 50%	**13.** 7%	**18.** 100%
4. 25%	**9.** 24%	**14.** 37%	**19.** 30%
5. 75%	**10.** 5%	**15.** 45%	**20.** 28%

Sometimes we are given a fractional percentage such as $4\frac{1}{2}\%$; then we need to write the given fraction as an improper fraction first,

i.e. as $\frac{9}{2}$

EXERCISE 10b

Give $2\frac{1}{2}\%$ as a fraction in its simplest terms.

$$2\frac{1}{2}\% = \frac{5}{2}\%$$
$$= \frac{5}{2 \times 100} = \frac{1}{40}$$

Percentages 123

Give the following percentages as fractions. Simplify each fraction if it is possible.

1. $7\frac{1}{2}\%$
2. $3\frac{1}{3}\%$
3. $3\frac{1}{2}\%$
4. $10\frac{1}{2}\%$
5. $12\frac{1}{2}\%$
6. $33\frac{1}{3}\%$
7. $6\frac{1}{4}\%$
8. $16\frac{2}{3}\%$
9. $32\frac{1}{2}\%$
10. $11\frac{1}{2}\%$
11. $8\frac{1}{3}\%$
12. $2\frac{3}{4}\%$

Percentages can give fractions which are greater than 1.

Express 450% as a fraction.

$$450\% = \frac{\cancel{450}^{9}}{\cancel{100}_{2}}$$
$$= \frac{9}{2}$$
$$= 4\frac{1}{2}$$

Find the following percentages as whole numbers or fractions.

13. 350%
14. 200%
15. 500%
16. 150%
17. 720%
18. 120%
19. 180%
20. 105%

In questions 21 to 24, several alternative answers are given. Write down the letter that corresponds to the correct answer.

21. 8% is

   A $\frac{1}{8}$ B $\frac{8}{10}$ C $\frac{2}{5}$ D $\frac{2}{25}$

22. 95% is

   A $9\frac{1}{2}$ B $\frac{95}{10}$ C $\frac{19}{20}$ D 95

23. 750% is

   A $\frac{75}{100}$ B $7\frac{1}{2}$ C 75   D $\frac{75}{1000}$

24. $4\frac{1}{2}\%$ is

   A $\frac{2}{9}$ B $\frac{9}{20}$ C $\frac{9}{200}$ D $\frac{2}{90}$

CHANGING PERCENTAGES TO DECIMALS

Sometimes we want to write a percentage as a decimal.

EXERCISE 10c

Express as decimals   a) 62%   b) 4%

a) $62\% = \frac{62}{100}$

$ = 0.62$

b) $4\% = \frac{4}{100}$

$ = 0.04$

Express the following percentages as decimals.

1. 43%
2. 55%
3. 29%
4. 3%
5. 63%
6. 84%
7. 5%
8. 10%
9. 6%
10. 12%
11. 2%
12. 15%

Express $5\frac{1}{2}\%$ as a decimal.

$5\frac{1}{2}\% = 5.5\%$

$\phantom{5\frac{1}{2}\%} = \frac{5.5}{100}$

$\phantom{5\frac{1}{2}\%} = 0.055$

Express the following percentages as decimals.

13. 4.5%
14. 12.5%
15. $7\frac{1}{2}\%$
16. 32.5%
17. $9\frac{1}{2}\%$
18. 43.2%
19. $15\frac{1}{2}\%$
20. 2.5%
21. $24\frac{1}{2}\%$

COMMON PERCENTAGES

It is worthwhile remembering that

25% is $\frac{1}{4}$, 50% is $\frac{1}{2}$ and 100% is 1.

PERCENTAGES OF QUANTITIES

Percentages are not much use on their own. We should always think of percentages *of something*. For example, a 5% rise in wages means nothing until we know what it is 5% of, e.g. 5% of £80.

EXERCISE 10d

Find 6% of 75 p.

$$6\% \text{ of } 75\,\text{p} = \frac{\cancel{6}^{\,3}}{\cancel{100}_{\,2}} \times \frac{\cancel{75}^{\,3}}{1}\,\text{p}$$

$$= \frac{9}{2}\,\text{p}$$

$$= 4\tfrac{1}{2}\,\text{p}$$

Find:

1. 5% of 40 p
2. 10% of 600 m
3. 15% of 60 m
4. 32% of £25
5. 75% of 16 cm
6. 20% of £5
7. 25% of 4 p
8. 12% of £100
9. 27% of 250 g
10. 36% of 75 m
11. 75% of 80 cm
12. 63% of 900 km
13. 35% of £80
14. 2% of 300 mm
15. 95% of 240 p

16. 40% of £15
17. 50% of 28 g
18. 80% of £45
19. 85% of 8000 kg
20. 4% of 200 km
21. 6% of 250 m
22. 20% of 12 km
23. 25% of 8 p
24. 50% of 18 p
25. 8% of £50
26. 1% of 4000 m
27. 32% of 50 kg
28. 86% of 450 p
29. 5% of 320 m
30. 16% of 125 m

> Find $4\frac{1}{2}\%$ of $350\,\text{cm}$
>
> $$4\frac{1}{2}\% = \frac{9}{2}\%$$
>
> $$4\frac{1}{2}\% \text{ of } 350\,\text{cm} = \frac{9}{2 \times \cancel{100}_2} \times \frac{\cancel{350}^7}{1}\,\text{cm}$$
>
> $$= \frac{63}{4}\,\text{cm}$$
>
> $$= 15.75\,\text{cm}$$

Find:

31. $2\frac{1}{2}\%$ of $80\,\text{p}$

32. $7\frac{1}{2}\%$ of $120\,\text{m}$

33. $12\frac{1}{2}\%$ of $32\,\text{kg}$

34. $37\frac{1}{2}\%$ of £88

35. $3\frac{1}{3}\%$ of $60\,\text{p}$

36. $16\frac{2}{3}\%$ of $12\,\text{m}$

37. $5\frac{1}{2}\%$ of $400\,\text{g}$

38. $1\frac{1}{2}\%$ of $600\,\text{m}$

39. $22\frac{1}{2}\%$ of $160\,\text{kg}$

40. $27\frac{1}{2}\%$ of £200

41. $17\frac{1}{2}\%$ of $240\,\text{g}$

42. $6\frac{1}{4}\%$ of $32\,\text{p}$

43. $33\frac{1}{3}\%$ of $66\,\text{m}$

44. $66\frac{2}{3}\%$ of $36\,\text{m}$

In questions 45 to 47, several alternative answers are given. Write down the letter that corresponds to the correct answer.

45. 10% of 50 m is

    A 500 m   **B** 5 m   **C** 10 m   **D** 25 m

46. 200% of 80 kg is

    A 40 kg   **B** 16 kg   **C** 160 kg   **D** 200 kg

47. 50% of 24 m is

    A 48 m   **B** 12 m   **C** 120 m   **D** 2.4 m

PROBLEMS

EXERCISE 10e

> Mr Sharp pays a deposit of 15% on a second-hand car whose cost is £900. How much is the deposit?
>
> The deposit is 15% of £900
>
> $= \frac{15}{100} \times £900$
>
> $= £\frac{15}{\cancel{100}} \times \cancel{900}^{9}$
>
> $= £135$

1. In a greengrocer's shop, a box contains 80 oranges. If 5% of the oranges are bad, how many bad oranges are there?

2. In a biology examination there were 70 candidates and 90% passed. How many pupils passed the examination?

3. During an epidemic, 24% of the pupils of a primary school caught chickenpox. If there were 300 pupils in the school, how many caught chickenpox?

4. In a local election, 32% of the electorate did not vote. There were 2500 on the electoral list. How many did not vote?

5. Seven years ago Mrs Allen bought a car that cost £1100. If she is now selling it for 35% of the cost price, what is her selling price?

6. Of the pupils in a school 49% are boys. There are 1200 pupils in the school altogether. How many boys are there?

7. Carol saves 30% of the money her grandmother gives her. For her birthday this year her grandmother has given her £2. How much does Carol save this time?

8. 72% of the vehicles passing the school gate are cars. If 250 vehicles pass the gate in one day, how many of them are cars?

9. Of the residents in a block of flats, 12% are under fifteen years of age. If 125 people live in the block, how many of them are under fifteen?

10. 62% of the population of a town are over 75 years old. The population is 12600. How many people over 75 years old live in the town?

FURTHER PROBLEMS

When we have 100% of a quantity we have the whole of the quantity. If 40% of the pupils in a class are boys then (100 − 40)% of the pupils are girls, i.e. 60% are girls.

EXERCISE 10f

> If Alan spends 82% of his weekly money, what percentage does he save?
>
> Alan saves (100 − 82)%
>
> = 18% of his money.

1. On Monday, 97% of the pupils are in school. What percentage of the pupils are absent?

2. When Mr Adams ordered a sofa, he paid a deposit of 15% of its price. What percentage of its price is there still to pay?

3. Philip is asleep for 28% of any 24 hour day. For what percentage of the day is he awake?

4. In a train to London, 93% of the seats are filled. What percentage of the seats are empty?

5. A family spends 54% of its income on food and 20% on clothes. What percentage is not spent on food or clothes?

6. In an election, Mr Garfield and Mrs Lincoln each received 48% of the votes and the remaining voting papers were spoiled. What percentage of the voting papers were spoiled?

7. Of a set of books, 6% are damaged in some way and cannot be used. What percentage of the books are usable?

8. Light bulbs made in a factory are tested and, in one batch, 8% are found to be faulty. What percentage are not faulty?

9. At a jumble sale, 64% of the jumble is sold during the sale, 24% is taken away afterwards by a dealer and the rest is rubbish. What percentage of the jumble is rubbish?

Percentages

CHANGING FRACTIONS TO PERCENTAGES

So far we have started with a percentage and turned it into a fraction to use it.

We sometimes need to work in the opposite direction and change fractions to percentages.

To change a fraction into a percentage we mutiply the fraction by 100%,

i.e.
$$\frac{7}{10} = \frac{7}{10} \times 100\%$$
$$= 70\%$$

EXERCISE 10g

> Express $\frac{3}{5}$ as a percentage.
> $$\frac{3}{5} = \frac{3}{5} \times 100\%$$
> $$= 60\%$$

Express the following fractions as percentages.

1. $\frac{1}{5}$
2. $\frac{1}{4}$
3. $\frac{7}{20}$
4. $\frac{3}{4}$
5. $\frac{1}{3}$
6. $\frac{4}{5}$
7. $\frac{3}{10}$
8. $\frac{1}{25}$
9. $\frac{7}{40}$
10. $\frac{9}{20}$
11. $\frac{3}{25}$
12. $\frac{63}{100}$
13. $\frac{1}{8}$
14. $\frac{5}{8}$
15. $\frac{9}{10}$
16. $\frac{91}{100}$

In questions 17 to 20, several alternative answers are given. Write down the letter that corresponds to the correct answer.

17. Written as a percentage, $\frac{7}{8}$ is

    A 7%   **B** 70%   **C** $87\frac{1}{2}\%$ **D** $92\frac{1}{2}\%$

18. Written as a percentage, $\frac{11}{20}$ is

    A 5.5%   **B** 11%   **C** 110%   **D** 55%

19. Written as a percentage, $\frac{9}{100}$ is

    A 9%   **B** 90%   **C** 0.09%   **D** 900%

20. Written as a percentage, $\frac{1}{50}$ is

    A 50%   **B** 2%   **C** 1%   **D** $\frac{1}{2}\%$

PERCENTAGES, FRACTIONS AND DECIMALS

EXERCISE 10h

1. Express $\frac{6}{25}$ as a) a percentage b) a decimal
2. Express 62% as a) a fraction b) a decimal
3. Express 48% as a) a decimal b) a fraction
4. Express 0.82 as a) a percentage b) a fraction

Copy and complete the following table. The working needed is similar to the working required for questions 1 to 4.

	Fraction	Percentage	Decimal
5.		50%	
6.	$\frac{3}{4}$		
7.			0.25
8.	$\frac{2}{5}$		
9.	$\frac{9}{10}$		
10.			0.7
11.		5%	

EXPRESSING ONE QUANTITY AS A FRACTION OF ANOTHER

EXERCISE 10i

> Give 4 m as a fraction of 24 m
>
> $4\,\text{m} = \frac{\cancel{4}^{1}}{\cancel{24}_{6}}$ of 24 m
>
> $= \frac{1}{6}$ of 24 m

1. Give 5 p as a fraction of 10 p.
2. Give 8 m as a fraction of 24 m.
3. Give 5 g as a fraction of 20 g.
4. Give 24 m as a fraction of 60 m.
5. Give 40 cm as a fraction of 45 cm.

In questions 6 to 9, express the first quantity as a fraction of the second.

6. £7 ; £28
7. 13 m ; 40 m
8. 32 p ; 50 p
9. 12 kg ; 72 kg

Percentages

EXPRESSING ONE QUANTITY AS A PERCENTAGE OF ANOTHER

EXERCISE 10j

> Give 3 m as a percentage of 12 m
>
> $$3\,m = \frac{3}{12} \text{ of } 12\,m$$
>
> $$= \frac{3}{12} \times \frac{100}{1} \% \text{ of } 12\,m$$
>
> $$= 25\% \text{ of } 12\,m$$

1. Give 2 p as a percentage of 10 p.
2. Give £15 as a percentage of £50.
3. Give 3 g as a percentage of 75 g.
4. Give 18 m as a percentage of 54 m.
5. Give 25 kg as a percentage of 125 kg.

Express the first quantity as a percentage of the second.

6. 14 p ; 56 p
7. 45 m ; 180 m
8. £6 ; £60
9. 11 cm ; 55 cm
10. 4 km ; 50 km
11. 8 hours ; 24 hours
12. 36 g ; 120 g
13. 65 miles ; 130 miles
14. 17 p ; 20 p

MIXED UNITS

Sometimes the two quantities are expressed in different units, so one of the units has to be changed. It is usually easier to work with the smaller unit. For example, if we are given centimetres and metres then we work in centimetres.

EXERCISE 10k

> Express 63 p as a percentage of £3
>
> $$£3 = 300\,p$$
>
> $$63\,p = \frac{63}{300} \times \frac{100}{1} \% \text{ of } £3$$
>
> $$= 21\% \text{ of } £3$$

1. Express 16 cm as a percentage of 4 m.
2. Express 40 p as a percentage of £2.
3. Express 4 cm as a percentage of 1 m.
4. Express 24 g as a percentage of 2 kg.
5. Express 60 m as a percentage of 3 km.

Express the first quantity as a percentage of the second.

6. 5 p ; £1
7. 150 cm ; 2 m
8. 40 m ; 2 km
9. 45 g ; 2 kg
10. 5 cm ; 80 mm
11. 680 m ; 5 km
12. 72 p ; £3.60
13. 4 m ; 200 cm
14. £1.26 ; 600 p

PROBLEMS

EXERCISE 10I

In a class of 25 pupils there are 14 girls. What percentage of the class are girls?

$$14 \text{ girls} = \frac{14}{25} \text{ of the class}$$

$$= \frac{14}{25} \times 100 \%$$

$$= 56\%$$

So 56% of the class are girls.

1. In a mathematics lesson, 12 pupils are writing in pencil. There are 25 pupils in the class. What percentage of them are writing in pencil?

2. In a cricket match, John made 17 runs and the total number of runs scored was 85. Express the number of runs scored by John as a percentage of the total.

3. The profit from selling a picture is £800. Mr Adams gets £300 of it and his wife gets £500. What percentage of the profit does Mr Adams receive and what percentage does his wife receive?

4. A man goes on a 50 mile journey. If he travels for 16 miles on a motorway, what percentage of the journey is this?

5. In a box containing 120 eggs, 6 are bad. What percentage of the eggs are good?

6. Mark's journey to school is 1500 m. He runs 120 m and walks the rest. What percentage of the whole distance does he run?

7. Andrew sleeps for eight hours of a twenty-four hour day. For what percentage of the day is he awake?

8. At the beginning of the day, a greengrocer's shop has 200 oranges. If 48 are sold during the day what percentage of the oranges are sold?

MIXED EXERCISES

EXERCISE 10m
1. Express 8% as   a) a fraction   b) a decimal.
2. Express $\frac{3}{5}$ as   a) a percentage   b) a decimal.
3. Find 45% of 40 p.
4. Express $5\frac{1}{2}$% as a fraction.
5. 65% of the vehicles passing along a major road are lorries. If 300 vehicles pass, how many of them are lorries?
6. If 8% of a box of ballpoint pens will not write, what percentage of the pens are usable?
7. In a box of 200 bananas, 14 bananas are bad. What percentage of the bananas are bad?

EXERCISE 10n
1. Express 26% as a fraction.
2. Express $\frac{13}{20}$ as a percentage.
3. Find 75% of £4.40.
4. Express 67% as a decimal.
5. In school, 9% of the time is spent on mathematics lessons. What percentage of the time is *not* spent on mathematics lessons?
6. The Martin family spend 20% of their holiday time in travelling. Their holiday lasts ten days. How long do they spend travelling?
7. The Martin family take 80 kg of luggage with them on holiday. Angela Martin takes 16 kg. What percentage is this of the total amount of luggage?
8. The Martin family take £500 spending money on holiday. They spend £400. What percentage of their spending money is left?

EXERCISE 10p In each question, several alternative answers are given. Write down the letter that corresponds to the correct answer.

1. Expressed as a fraction, 30% is

   A $\frac{1}{3}$ **B** $2\frac{1}{30}$ **C** $\frac{3}{10}$ **D** 30

2. Expressed as a percentage, $\frac{1}{5}$ is

   A 2%   **B** 5%   **C** 20%   **D** 50%

3. 16% of 750 cm is

   A 16 cm   **B** 75 cm   **C** 120 cm   **D** 160 cm

4. In a school election, 30% vote for Mary and 40% vote for Anne. The percentage who vote for neither Mary nor Anne is

   A 70%   **B** 10%   **C** 40%   **D** 30%

5. Gavin had 70 p in his pocket. He spent 7 p. The percentage of his money that he spent was

   A 50%   **B** 7%   **C** 10%   **D** 70%

6. In one innings of a cricket match Jack Shore made 50% of the total number of runs. Which of the following statements must be true?

   A The team made 100 runs.

   B Jack Shore made half the runs.

   C Jack Shore made 50 runs.

   D Only two people batted.

11 MATRICES

SHOPPING LISTS

Mrs Smith and Mrs Jones go shopping for oranges and grapefruit.
Mrs Smith buys 6 oranges and 2 grapefruit.
Mrs Jones buys 5 oranges and 1 grapefruit.

We can arrange this information in a table.

	Oranges	Grapefruit
Mrs Smith	6	2
Mrs Jones	5	1

We can write this information more briefly in the form

$$\begin{pmatrix} 6 & 2 \\ 5 & 1 \end{pmatrix}$$

All the numbers are given, though we have missed out the words.

$\begin{pmatrix} 6 & 2 \\ 5 & 1 \end{pmatrix}$ is called a *matrix* (the plural is *matrices*).

It is held together with curved brackets.
Each number is called an *entry*.

A matrix can have any number of *rows* and *columns*; rows go across and columns go down.

$\begin{pmatrix} 4 & 2 & 3 \\ 5 & 6 & 1 \end{pmatrix}$ is a matrix. $\begin{pmatrix} 3 & 2 \\ 1 & 4 \\ 0 & 5 \end{pmatrix}$ and $(1 \ \ 2)$ also are matrices.

SIZE OF A MATRIX

We count the number of rows and columns and use these to describe the size of the matrix. The number of rows comes first, then the number of columns.

$\begin{pmatrix} 6 & 2 & 3 \\ 5 & 1 & 2 \end{pmatrix}$ is a 2 × 3 matrix and $\begin{pmatrix} 6 & 3 \\ 0 & 1 \\ 4 & 3 \end{pmatrix}$ is a 3 × 2 matrix.

EXERCISE 11a

> Give the sizes of the matrices $(1 \quad 4 \quad 1)$ and $\begin{pmatrix} 2 & 4 \\ 3 & 6 \end{pmatrix}$
>
> $(1 \quad 4 \quad 1)$ is a 1×3 matrix
>
> $\begin{pmatrix} 2 & 4 \\ 3 & 6 \end{pmatrix}$ is a 2×2 matrix

Give the sizes of the following matrices.

1. $\begin{pmatrix} 4 & 1 \\ 2 & 3 \end{pmatrix}$

2. $\begin{pmatrix} 2 \\ 1 \end{pmatrix}$

3. $\begin{pmatrix} 3 & 1 \\ 1 & 4 \\ 0 & 6 \end{pmatrix}$

4. $\begin{pmatrix} 2 & 6 & 8 \\ 3 & 1 & 2 \end{pmatrix}$

5. $(1 \quad 6)$

6. $\begin{pmatrix} 5 \\ 4 \\ 2 \end{pmatrix}$

7. $(1 \quad 3 \quad 2)$

8. $\begin{pmatrix} 3 & 4 \\ 2 & 7 \end{pmatrix}$

9. $\begin{pmatrix} 3 & 4 & 6 \\ 2 & 1 & 0 \\ 4 & 3 & 1 \end{pmatrix}$

> Give a) the second row b) the third column
>
> in the matrix $\begin{pmatrix} 1 & 4 & 3 \\ 2 & 1 & 8 \end{pmatrix}$
>
> a) The second row is $2 \quad 1 \quad 8$
>
> b) The third column is $\begin{matrix} 3 \\ 8 \end{matrix}$

10. Give the second row in the matrix $\begin{pmatrix} 4 & 2 \\ 1 & 4 \end{pmatrix}$

11. Give the first row in the matrix $\begin{pmatrix} 4 \\ 1 \end{pmatrix}$

12. Give the first column in the matrix $\begin{pmatrix} 4 & 1 & 1 \\ 3 & 2 & 4 \\ 2 & 5 & 8 \end{pmatrix}$

13. Give the second column in the matrix $\begin{pmatrix} 1 & 3 & 4 \\ 2 & 0 & 8 \end{pmatrix}$

14. Give a) the first row   b) the first column

in the matrix $\begin{pmatrix} 7 & 3 \\ 1 & 4 \end{pmatrix}$

15. Give a) the third row   b) the second column

in the matrix $\begin{pmatrix} 4 & 8 \\ 9 & 0 \\ 6 & 1 \end{pmatrix}$

16. Give a) the first row   b) the second column
in the matrix $(4 \quad 8 \quad 1 \quad 3)$

Give the entry in the second row of the first column

of the matrix $\begin{pmatrix} 4 & 8 \\ 3 & 1 \end{pmatrix}$

$\begin{pmatrix} 4 & 8 \\ ③ & 1 \end{pmatrix}$ The entry is 3.

---

17. Give the entry in the first row of the first column

of the matrix $\begin{pmatrix} 1 & 8 \\ 9 & 2 \end{pmatrix}$

18. Give the entry in the second row of the first column

of the matrix $\begin{pmatrix} 4 & 6 \\ 2 & 7 \end{pmatrix}$

19. Give the entry in the first row of the second column
of the matrix $(4 \quad 3 \quad 6)$

20. Give the entry in the third row of the first column

of the matrix $\begin{pmatrix} 4 \\ 5 \\ 3 \end{pmatrix}$

In questions 21 to 23, several alternative answers are given. Write down the letter that corresponds to the correct answer.

21. The size of the matrix $\begin{pmatrix} 4 \\ 3 \end{pmatrix}$ is

    A 2×1 **B** 1×2 **C** 4×3

22. The size of the matrix $\begin{pmatrix} 4 & 1 & 4 \end{pmatrix}$ is

    A 3×1 **B** 4×1 **C** 1×3

23. The entry in the first row of the second column in the matrix $\begin{pmatrix} 4 & 1 \\ 3 & 2 \end{pmatrix}$ is

    A 4   **B** 1   **C** 3   **D** 2

ADDITION OF MATRICES

Mrs Smith's and Mrs Jones' first shopping lists were represented by the matrix $\begin{pmatrix} 6 & 2 \\ 5 & 1 \end{pmatrix}$

The next week, Mrs Smith buys 3 oranges and no grapefruit and Mrs Jones buys 4 oranges and 2 grapefruit. We can arrange this information in a table:

	Oranges	Grapefruit
Mrs Smith	3	0
Mrs Jones	4	2

or as a matrix $\begin{pmatrix} 3 & 0 \\ 4 & 2 \end{pmatrix}$

By adding the amount of fruit bought each week we see that, over the two weeks, Mrs Smith buys 9 oranges and 2 grapefruit while Mrs Jones buys 9 oranges and 3 grapefruit.

This can be written in the form

$$\begin{pmatrix} 6 & 2 \\ 5 & 1 \end{pmatrix} + \begin{pmatrix} 3 & 0 \\ 4 & 2 \end{pmatrix} = \begin{pmatrix} 6+3 & 2+0 \\ 5+4 & 1+2 \end{pmatrix}$$

$$= \begin{pmatrix} 9 & 2 \\ 9 & 3 \end{pmatrix}$$

Notice that we add the entries in the corresponding positions in the two matrices. The final matrix is the same size as the two given matrices.

EXERCISE 11b

Find a) $\begin{pmatrix} 1 \\ 4 \end{pmatrix} + \begin{pmatrix} 3 \\ 6 \end{pmatrix}$ b) $\begin{pmatrix} 3 & 5 \\ 4 & 1 \end{pmatrix} + \begin{pmatrix} 0 & 5 \\ 3 & 4 \end{pmatrix}$

a) $\begin{pmatrix} 1 \\ 4 \end{pmatrix} + \begin{pmatrix} 3 \\ 6 \end{pmatrix} = \begin{pmatrix} 1+3 \\ 4+6 \end{pmatrix} = \begin{pmatrix} 4 \\ 10 \end{pmatrix}$

b) $\begin{pmatrix} 3 & 5 \\ 4 & 1 \end{pmatrix} + \begin{pmatrix} 0 & 5 \\ 3 & 4 \end{pmatrix} = \begin{pmatrix} 3+0 & 5+5 \\ 4+3 & 1+4 \end{pmatrix} = \begin{pmatrix} 3 & 10 \\ 7 & 5 \end{pmatrix}$

Find:

1. $\begin{pmatrix} 3 & 1 \\ 4 & 2 \end{pmatrix} + \begin{pmatrix} 2 & 6 \\ 8 & 0 \end{pmatrix}$

2. $\begin{pmatrix} 4 \\ 2 \end{pmatrix} + \begin{pmatrix} 3 \\ 5 \end{pmatrix}$

3. $(3 \quad 4) + (1 \quad 2)$

4. $\begin{pmatrix} 4 & 5 \\ 6 & 1 \end{pmatrix} + \begin{pmatrix} 2 & 3 \\ 0 & 4 \end{pmatrix}$

5. $\begin{pmatrix} 7 & 1 \\ 4 & 1 \end{pmatrix} + \begin{pmatrix} 3 & 2 \\ 3 & 3 \end{pmatrix}$

6. $\begin{pmatrix} 4 & 3 \\ 2 & 1 \end{pmatrix} + \begin{pmatrix} 6 & 1 \\ 3 & 0 \end{pmatrix}$

7. $\begin{pmatrix} 4 \\ 2 \end{pmatrix} + \begin{pmatrix} 8 \\ 4 \end{pmatrix}$

8. $(1 \quad 2) + (3 \quad 4)$

9. $\begin{pmatrix} 3 & 5 \\ 5 & 3 \end{pmatrix} + \begin{pmatrix} 9 & 0 \\ 1 & 2 \end{pmatrix}$

10. $\begin{pmatrix} 4 & 8 \\ 6 & 4 \end{pmatrix} + \begin{pmatrix} 10 & 11 \\ 3 & 7 \end{pmatrix}$

11. $\begin{pmatrix} 1 & 4 & 8 \\ 1 & 2 & 1 \end{pmatrix} + \begin{pmatrix} 4 & 5 & 1 \\ 3 & 8 & 1 \end{pmatrix}$

13. $\begin{pmatrix} 1 & 4 \\ 3 & 8 \\ 0 & 1 \end{pmatrix} + \begin{pmatrix} 6 & 2 \\ 1 & 3 \\ 4 & 8 \end{pmatrix}$

12. $(3 \quad 4 \quad 2 \quad 0) + (3 \quad 2 \quad 1 \quad 5)$

14. $\begin{pmatrix} 1 \\ 2 \\ 3 \end{pmatrix} + \begin{pmatrix} 4 \\ 5 \\ 6 \end{pmatrix}$

In questions 15 to 18, several alternative answers are given. Write down the letter that corresponds to the correct answer.

15. $\begin{pmatrix} 4 & 2 \\ 3 & 1 \end{pmatrix} + \begin{pmatrix} 6 & 8 \\ 1 & 3 \end{pmatrix}$ is equal to

A $\begin{pmatrix} 12 & 10 \\ 9 & 2 \end{pmatrix}$ **B** $\begin{pmatrix} 10 & 16 \\ 4 & 4 \end{pmatrix}$ **C** $\begin{pmatrix} 24 & 16 \\ 3 & 3 \end{pmatrix}$ **D** None of these

16. $\begin{pmatrix} 4 \\ 1 \end{pmatrix} + \begin{pmatrix} 3 \\ 5 \end{pmatrix}$ is equal to

A $\begin{pmatrix} 7 \\ 6 \end{pmatrix}$ **B** $\begin{pmatrix} 12 \\ 5 \end{pmatrix}$ **C** $\begin{pmatrix} 7 \\ 5 \end{pmatrix}$ **D** None of these

17. $(3 \quad 2) + (4 \quad 1)$ is equal to

A $(4 \quad 6)$ **B** $(6 \quad 4)$ **C** $(7 \quad 3)$ **D** $(3 \quad 7)$

18. $\begin{pmatrix} 4 & 2 \\ 0 & 3 \end{pmatrix} + \begin{pmatrix} 7 & 1 \\ 5 & 0 \end{pmatrix}$ is equal to

A $\begin{pmatrix} 11 & 3 \\ 0 & 0 \end{pmatrix}$ **B** $\begin{pmatrix} 11 & 3 \\ 8 & 0 \end{pmatrix}$ **C** $\begin{pmatrix} 11 & 3 \\ 5 & 3 \end{pmatrix}$ **D** None of these

We can add matrices only if they are of the same size.
(In Exercise 11b each pair of matrices that were to be added were the same size.)

If matrices are of different sizes they cannot be added.

$\begin{pmatrix} 1 & 2 \\ 3 & 4 \end{pmatrix}$ and $\begin{pmatrix} 3 & 4 & 8 \\ 1 & 0 & 5 \end{pmatrix}$ cannot be added, nor can $(1 \quad 3)$ and $\begin{pmatrix} 4 \\ 6 \end{pmatrix}$

EXERCISE 11c

a) Write down the sizes of the matrices

$$\begin{pmatrix} 1 & 4 \\ 3 & 4 \end{pmatrix}, \begin{pmatrix} 4 & 0 \\ 8 & 4 \\ 1 & 3 \end{pmatrix} \text{ and } \begin{pmatrix} 2 & 3 \\ 1 & 2 \end{pmatrix}$$

b) Is it possible to add $\begin{pmatrix} 1 & 4 \\ 3 & 2 \end{pmatrix}$ and $\begin{pmatrix} 4 & 0 \\ 8 & 4 \\ 1 & 3 \end{pmatrix}$

c) Is it possible to add $\begin{pmatrix} 1 & 4 \\ 3 & 2 \end{pmatrix}$ and $\begin{pmatrix} 2 & 3 \\ 1 & 2 \end{pmatrix}$

a) $\begin{pmatrix} 1 & 4 \\ 3 & 2 \end{pmatrix}$ is a 2 × 2 matrix $\begin{pmatrix} 4 & 0 \\ 8 & 4 \\ 1 & 3 \end{pmatrix}$ is a 3 × 2 matrix

$\begin{pmatrix} 2 & 3 \\ 1 & 2 \end{pmatrix}$ is a 2 × 2 matrix

b) No, because the two matrices are of different sizes.

c) Yes, because the two matrices are the same size.

Write down the sizes of the matrices in questions 1 to 10. In each case state whether or not the pair of matrices can be added.

1. $\begin{pmatrix} 2 & 3 \\ 1 & 0 \end{pmatrix}$ and $\begin{pmatrix} 5 & 3 \\ 2 & 4 \end{pmatrix}$

2. $(2 \ 3)$ and $\begin{pmatrix} 4 \\ 3 \end{pmatrix}$

3. $(2 \ 4)$ and $(5 \ 3)$

4. $\begin{pmatrix} 3 & 4 \\ 5 & 6 \end{pmatrix}$ and $\begin{pmatrix} 5 \\ 0 \end{pmatrix}$

5. $(3 \ 4)$ and $\begin{pmatrix} 2 & 5 \\ 1 & 0 \end{pmatrix}$

6. $\begin{pmatrix} 1 \\ 2 \\ 3 \end{pmatrix}$ and $\begin{pmatrix} 6 \\ 5 \end{pmatrix}$

7. $\begin{pmatrix} 3 & 0 \\ 5 & 3 \end{pmatrix}$ and $\begin{pmatrix} 4 & 1 \\ 3 & 2 \end{pmatrix}$

8. $\begin{pmatrix} 3 \\ 1 \end{pmatrix}$ and $\begin{pmatrix} 5 \\ 4 \end{pmatrix}$

9. $\begin{pmatrix} 6 \\ 7 \end{pmatrix}$ and $(4 \ 5)$

10. $\begin{pmatrix} 3 & 4 \\ 1 & 4 \end{pmatrix}$ and $\begin{pmatrix} 2 & 4 & 1 \\ 1 & 2 & 3 \end{pmatrix}$

In each of the questions 11 to 18, write down the sizes of the two matrices. If they are the same size, add them. If they are not the same size write "They cannot be added".

11. $\begin{pmatrix} 2 & 4 \\ 3 & 2 \end{pmatrix}$ and $\begin{pmatrix} 1 & 4 \\ 3 & 6 \end{pmatrix}$

12. $\begin{pmatrix} 4 \\ 1 \end{pmatrix}$ and $\begin{pmatrix} 5 \\ 6 \end{pmatrix}$

13. $(5 \quad 3)$ and $(3 \quad 5)$

14. $(3 \quad 5)$ and $\begin{pmatrix} 3 \\ 1 \end{pmatrix}$

15. $(3 \quad 4 \quad 2)$ and $(4 \quad 1 \quad 2)$

16. $\begin{pmatrix} 7 & 8 \\ 2 & 8 \end{pmatrix}$ and $\begin{pmatrix} 0 & 3 \\ 0 & 2 \end{pmatrix}$

17. $(3 \quad 5)$ and $(4 \quad 3)$

18. $\begin{pmatrix} 4 & 3 \\ 5 & 1 \end{pmatrix}$ and $\begin{pmatrix} 3 \\ 3 \end{pmatrix}$

Find where it is possible:

19. $\begin{pmatrix} 9 & 1 \\ 4 & 2 \end{pmatrix} + \begin{pmatrix} 3 & 3 \\ 1 & 1 \end{pmatrix}$

20. $(3 \quad 5) + (2 \quad 4)$

21. $(4 \quad 6) + (7 \quad 2 \quad 9)$

22. $\begin{pmatrix} 4 & 7 \\ 3 & 1 \end{pmatrix} + \begin{pmatrix} 2 & 3 \\ 3 & 2 \end{pmatrix}$

23. $\begin{pmatrix} 4 & 3 \\ 5 & 0 \end{pmatrix} + (6 \quad 7)$

24. $\begin{pmatrix} 3 \\ 4 \end{pmatrix} + \begin{pmatrix} 6 \\ 1 \end{pmatrix}$

25. $(3 \quad 5 \quad 6) + (2 \quad 4 \quad 1)$

26. $\begin{pmatrix} 6 & 8 \\ 0 & 4 \end{pmatrix} + \begin{pmatrix} 3 & 5 \\ 0 & 5 \end{pmatrix}$

27. $\begin{pmatrix} 3 & 7 \\ 3 & 0 \end{pmatrix} + \begin{pmatrix} 1 & 0 \\ 3 & 1 \end{pmatrix}$

28. $(3 \quad 5) + \begin{pmatrix} 5 \\ 6 \end{pmatrix}$

SUBTRACTION OF MATRICES

We can subtract matrices in a similar way.

EXERCISE 11d

Find $\begin{pmatrix} 3 & 5 \\ 7 & 6 \end{pmatrix} - \begin{pmatrix} 1 & 2 \\ 1 & 0 \end{pmatrix}$

$\begin{pmatrix} 3 & 5 \\ 7 & 6 \end{pmatrix} - \begin{pmatrix} 1 & 2 \\ 1 & 0 \end{pmatrix} = \begin{pmatrix} 2 & 3 \\ 6 & 6 \end{pmatrix}$

Find:

1. $\begin{pmatrix} 3 & 7 \\ 8 & 4 \end{pmatrix} - \begin{pmatrix} 2 & 4 \\ 1 & 0 \end{pmatrix}$

2. $\begin{pmatrix} 4 \\ 3 \end{pmatrix} - \begin{pmatrix} 3 \\ 1 \end{pmatrix}$

3. $\begin{pmatrix} 7 & 9 \\ 6 & 10 \end{pmatrix} - \begin{pmatrix} 3 & 4 \\ 5 & 9 \end{pmatrix}$

4. $\begin{pmatrix} 9 & 8 \\ 10 & 1 \end{pmatrix} - \begin{pmatrix} 4 & 0 \\ 3 & 0 \end{pmatrix}$

5. $(7 \ 8) - (3 \ 4)$

6. $\begin{pmatrix} 7 & 8 \\ 9 & 4 \end{pmatrix} - \begin{pmatrix} 6 & 3 \\ 1 & 1 \end{pmatrix}$

7. $\begin{pmatrix} 7 & 7 \\ 7 & 7 \end{pmatrix} - \begin{pmatrix} 3 & 4 \\ 2 & 1 \end{pmatrix}$

8. $\begin{pmatrix} 5 \\ 4 \end{pmatrix} - \begin{pmatrix} 3 \\ 1 \end{pmatrix}$

9. $\begin{pmatrix} 3 & 6 \\ 6 & 8 \end{pmatrix} - \begin{pmatrix} 1 & 0 \\ 6 & 8 \end{pmatrix}$

10. $(1 \ 6 \ 7) - (0 \ 3 \ 2)$

In questions 11 and 12, several alternative answers are given. Write down the letter that corresponds to the correct answer.

11. $\begin{pmatrix} 5 & 6 \\ 6 & 7 \end{pmatrix} - \begin{pmatrix} 3 & 4 \\ 1 & 0 \end{pmatrix}$ is equal to

A $\begin{pmatrix} 2 & 2 \\ 8 & 7 \end{pmatrix}$ **B** $\begin{pmatrix} 8 & 2 \\ 6 & 7 \end{pmatrix}$ **C** $\begin{pmatrix} 2 & 2 \\ 5 & 7 \end{pmatrix}$ **D** $\begin{pmatrix} 8 & 10 \\ 8 & 7 \end{pmatrix}$

12. $\begin{pmatrix} 4 \\ 9 \end{pmatrix} - \begin{pmatrix} 3 \\ 7 \end{pmatrix}$ is equal to

A $\begin{pmatrix} 7 \\ 16 \end{pmatrix}$ **B** $\begin{pmatrix} 1 \\ 16 \end{pmatrix}$ **C** $\begin{pmatrix} 1 \\ 2 \end{pmatrix}$ **D** $\begin{pmatrix} 7 \\ 2 \end{pmatrix}$

Matrices must be the same size as one another if we are to be able to subtract them.

In each of the questions 13 to 18, write down the sizes of the two matrices and say whether or not it is possible to subtract the second matrix from the first.

13. $\begin{pmatrix} 2 & 4 \\ 5 & 7 \end{pmatrix}$ and $\begin{pmatrix} 1 & 0 \\ 1 & 1 \end{pmatrix}$

14. $\begin{pmatrix} 2 & 5 \\ 3 & 0 \end{pmatrix}$ and $(1 \ 2)$

15. $\begin{pmatrix} 6 & 7 \\ 5 & 9 \end{pmatrix}$ and $\begin{pmatrix} 3 \\ 2 \end{pmatrix}$

16. $(4 \ 3)$ and $(3 \ 0)$

17. $\begin{pmatrix} 3 & 9 \\ 6 & 7 \end{pmatrix}$ and $\begin{pmatrix} 0 & 2 \\ 1 & 1 \end{pmatrix}$

18. $\begin{pmatrix} 4 \\ 3 \end{pmatrix}$ and $(3 \ 1 \ 0)$

In each of the questions 19 to 30, subtract the second matrix from the first if it is possible. If it is not possible, write "Subtraction cannot be done".

19. $\begin{pmatrix} 3 & 8 \\ 6 & 4 \end{pmatrix} - \begin{pmatrix} 1 & 7 \\ 0 & 2 \end{pmatrix}$

20. $\begin{pmatrix} 7 & 5 \\ 5 & 9 \end{pmatrix} - \begin{pmatrix} 3 & 4 \\ 3 & 6 \end{pmatrix}$

21. $\begin{pmatrix} 4 & 5 \\ 7 & 5 \end{pmatrix} - \begin{pmatrix} 2 & 3 \end{pmatrix}$

22. $\begin{pmatrix} 7 \\ 6 \end{pmatrix} - \begin{pmatrix} 3 & 4 \\ 1 & 0 \end{pmatrix}$

23. $\begin{pmatrix} 7 \\ 6 \end{pmatrix} - \begin{pmatrix} 3 \\ 1 \end{pmatrix}$

24. $(4 \ 5) - (2 \ 3)$

25. $\begin{pmatrix} 4 & 5 \\ 9 & 5 \end{pmatrix} - \begin{pmatrix} 0 & 2 \\ 9 & 2 \end{pmatrix}$

26. $\begin{pmatrix} 4 & 5 \\ 9 & 5 \end{pmatrix} - (0 \ 2)$

27. $\begin{pmatrix} 6 & 9 \\ 3 & 5 \end{pmatrix} - \begin{pmatrix} 5 & 6 \\ 2 & 5 \end{pmatrix}$

28. $\begin{pmatrix} 9 \\ 8 \end{pmatrix} - \begin{pmatrix} 4 & 5 \\ 3 & 2 \end{pmatrix}$

29. $\begin{pmatrix} 6 \\ 7 \\ 5 \end{pmatrix} - \begin{pmatrix} 5 \\ 6 \\ 4 \end{pmatrix}$

30. $(4 \ 5) - (0 \ 2)$

MULTIPLES OF MATRICES

If Mrs Smith and Mrs Jones each have the same shopping list for three weeks running we can see that

$$\begin{pmatrix} 3 & 1 \\ 1 & 2 \end{pmatrix} + \begin{pmatrix} 3 & 1 \\ 1 & 2 \end{pmatrix} + \begin{pmatrix} 3 & 1 \\ 1 & 2 \end{pmatrix} = 3\begin{pmatrix} 3 & 1 \\ 1 & 2 \end{pmatrix} = \begin{pmatrix} 9 & 3 \\ 3 & 6 \end{pmatrix}$$

In the same way, $5\begin{pmatrix} 1 & 4 \\ 3 & 2 \end{pmatrix} = \begin{pmatrix} 5 & 20 \\ 15 & 10 \end{pmatrix}$

When we multiply a matrix by 5 we multiply *every* entry by 5

EXERCISE 11e

Find $6\begin{pmatrix} 2 & 1 \\ 3 & 4 \end{pmatrix}$

$6\begin{pmatrix} 2 & 1 \\ 3 & 4 \end{pmatrix} = \begin{pmatrix} 12 & 6 \\ 18 & 24 \end{pmatrix}$

Find the following multiples.

1. $2\begin{pmatrix} 3 & 4 \\ 2 & 6 \end{pmatrix}$

2. $3\begin{pmatrix} 5 & 2 \\ 3 & 1 \end{pmatrix}$

3. $10\begin{pmatrix} 4 & 2 \\ 5 & 1 \end{pmatrix}$

4. $4(9 \quad 2)$

5. $2\begin{pmatrix} 9 \\ 4 \end{pmatrix}$

6. $5(5 \quad 3)$

7. $8\begin{pmatrix} 2 & 1 \\ 3 & 0 \end{pmatrix}$

8. $3\begin{pmatrix} 6 \\ 1 \end{pmatrix}$

9. $3\begin{pmatrix} 6 & 4 \\ 7 & 0 \end{pmatrix}$

MIXED QUESTIONS

EXERCISE 11f Find:

1. $\begin{pmatrix} 4 & 5 \\ 6 & 7 \end{pmatrix} - \begin{pmatrix} 2 & 1 \\ 0 & 1 \end{pmatrix}$

2. $\begin{pmatrix} 4 & 5 \\ 6 & 7 \end{pmatrix} + \begin{pmatrix} 2 & 1 \\ 0 & 1 \end{pmatrix}$

3. $(4 \quad 5) + (2 \quad 1)$

4. $6\begin{pmatrix} 4 & 5 \\ 6 & 7 \end{pmatrix}$

5. $\begin{pmatrix} 9 \\ 6 \end{pmatrix} + \begin{pmatrix} 3 \\ 4 \end{pmatrix}$

6. $\begin{pmatrix} 9 & 5 \\ 6 & 7 \end{pmatrix} - \begin{pmatrix} 3 & 5 \\ 4 & 3 \end{pmatrix}$

7. $\begin{pmatrix} 9 & 5 \\ 6 & 7 \end{pmatrix} + \begin{pmatrix} 3 & 5 \\ 4 & 3 \end{pmatrix}$

8. $3(2 \quad 1)$

9. $3\begin{pmatrix} 3 & 5 \\ 4 & 3 \end{pmatrix}$

10. $5\begin{pmatrix} 3 \\ 4 \end{pmatrix}$

11. $\begin{pmatrix} 10 & 7 \\ 9 & 6 \end{pmatrix} + \begin{pmatrix} 2 & 3 \\ 0 & 1 \end{pmatrix}$

12. $\begin{pmatrix} 10 & 7 \\ 9 & 6 \end{pmatrix} - \begin{pmatrix} 2 & 3 \\ 0 & 1 \end{pmatrix}$

13. $(10 \quad 7) + (2 \quad 3)$

14. $5\begin{pmatrix} 2 & 3 \\ 0 & 1 \end{pmatrix}$

15. $8\begin{pmatrix} 7 \\ 6 \end{pmatrix}$

16. $\begin{pmatrix} 2 & 3 & 4 \\ 2 & 5 & 3 \end{pmatrix} + \begin{pmatrix} 4 & 5 & 3 \\ 1 & 0 & 7 \end{pmatrix}$

17. $\begin{pmatrix} 5 \\ 6 \\ 9 \end{pmatrix} - \begin{pmatrix} 4 \\ 4 \\ 3 \end{pmatrix}$

18. $8(1 \quad 2 \quad 4)$

19. $\begin{pmatrix} 1 & 2 \\ 5 & 4 \end{pmatrix} + \begin{pmatrix} 4 & 5 \\ 1 & 2 \end{pmatrix} + \begin{pmatrix} 2 & 3 \\ 3 & 4 \end{pmatrix}$

20. $(3 \quad 5) - (1 \quad 0)$

In questions 21 to 23, several alternative answers are given. Write down the letter that corresponds to the correct answer.

21. $\begin{pmatrix} 2 & 4 \\ 3 & 5 \end{pmatrix} + \begin{pmatrix} 5 & 6 \\ 2 & 3 \end{pmatrix}$ is equal to

A $\begin{pmatrix} 6 & 11 \\ 8 & 5 \end{pmatrix}$ **B** $\begin{pmatrix} 7 & 10 \\ 5 & 8 \end{pmatrix}$ **C** $\begin{pmatrix} 2 & 4 & 5 & 6 \\ 3 & 5 & 2 & 3 \end{pmatrix}$ **D** $\begin{pmatrix} 17 \\ 13 \end{pmatrix}$

22. $(5 \ 6) - (3 \ 2)$ is equal to

A $(2 \ 4)$ **B** $(2 \ 3)$ **C** $(3 \ 4)$ **D** None of these

23. $3\begin{pmatrix} 5 & 4 \\ 3 & 2 \end{pmatrix}$ is equal to

A $\begin{pmatrix} 8 & 7 \\ 9 & 6 \end{pmatrix}$ **B** $\begin{pmatrix} 15 & 12 \\ 9 & 6 \end{pmatrix}$ **C** $\begin{pmatrix} 8 & 7 \\ 6 & 5 \end{pmatrix}$ **D** None of these

MATRICES WITH NEGATIVE ENTRIES

So far, all the entries in the matrices have been positive numbers but they can be negative as, for example, in $\begin{pmatrix} -2 & 4 \\ 1 & -6 \end{pmatrix}$

These matrices can still be added and subtracted and multiples can be found.

The next exercise starts with a reminder on addition and subtraction of negative numbers.

EXERCISE 11g

Find   a) $6 + (-4)$ c) $7 - (-3)$
        b) $(-8) + (-6)$ d) $-5 - (-1)$

a) $6 + (-4) = 6 - 4 = 2$
b) $(-8) + (-6) = -8 - 6 = -14$
c) $7 - (-3) = 7 + 3 = 10$
d) $-5 - (-1) = -5 + 1 = -4$

Find:

1. $4 + (-2)$
2. $-5 + 6$
3. $-7 + (-1)$
4. $-2 + (-3)$
5. $6 - (-2)$
6. $4 - (-3)$
7. $-5 - (-3)$
8. $-5 - 4$
9. $3 + (-4)$
10. $-7 - (-3)$
11. $-5 + 8$
12. $0 - (-4)$

Find $\begin{pmatrix} 6 & -2 \\ 3 & 0 \end{pmatrix} + \begin{pmatrix} 2 & 8 \\ -4 & 2 \end{pmatrix}$

$\begin{pmatrix} 6 & -2 \\ -3 & 0 \end{pmatrix} + \begin{pmatrix} 2 & 8 \\ -4 & 2 \end{pmatrix} = \begin{pmatrix} 6+2 & -2+8 \\ -3+(-4) & 0+2 \end{pmatrix}$

$= \begin{pmatrix} 8 & 6 \\ -7 & 2 \end{pmatrix}$

Find:

13. $\begin{pmatrix} 1 & 8 \\ 3 & 2 \end{pmatrix} + \begin{pmatrix} -3 & 6 \\ -2 & 1 \end{pmatrix}$

14. $(-3 \quad -5) + (8 \quad 9)$

15. $\begin{pmatrix} 1 \\ -4 \end{pmatrix} + \begin{pmatrix} -3 \\ -3 \end{pmatrix}$

16. $\begin{pmatrix} 4 & 1 \\ -3 & -1 \end{pmatrix} + \begin{pmatrix} 2 & -1 \\ 5 & 2 \end{pmatrix}$

17. $\begin{pmatrix} 6 & 0 \\ -3 & 2 \end{pmatrix} + \begin{pmatrix} -4 & 8 \\ 2 & 1 \end{pmatrix}$

18. $\begin{pmatrix} 4 & -3 \\ 3 & 1 \end{pmatrix} + \begin{pmatrix} 5 & 6 \\ 7 & -4 \end{pmatrix}$

19. $(2 \quad -1 \quad 3) + (3 \quad 4 \quad -2)$

20. $\begin{pmatrix} 4 \\ 3 \end{pmatrix} + \begin{pmatrix} -4 \\ 6 \end{pmatrix}$

21. $\begin{pmatrix} -5 & -3 \\ -7 & 9 \end{pmatrix} + \begin{pmatrix} 8 & 3 \\ -1 & 3 \end{pmatrix}$

22. $\begin{pmatrix} 3 & 5 \\ 4 & 2 \end{pmatrix} + \begin{pmatrix} -6 & -3 \\ -1 & 4 \end{pmatrix}$

Find $\begin{pmatrix} 5 & -4 \\ 6 & -1 \end{pmatrix} - \begin{pmatrix} 3 & 2 \\ -4 & -2 \end{pmatrix}$

$$\begin{pmatrix} 5 & -4 \\ 6 & -1 \end{pmatrix} - \begin{pmatrix} 3 & 2 \\ -4 & -2 \end{pmatrix} = \begin{pmatrix} 5-3 & -4-2 \\ 6-(-4) & -1-(-2) \end{pmatrix}$$
$$= \begin{pmatrix} 2 & -6 \\ 10 & 1 \end{pmatrix}$$

Find:

23. $\begin{pmatrix} 4 & 5 \\ -3 & 6 \end{pmatrix} - \begin{pmatrix} 5 & 8 \\ 4 & 2 \end{pmatrix}$

25. $\begin{pmatrix} 3 \\ 4 \\ 1 \end{pmatrix} - \begin{pmatrix} 1 \\ 3 \\ 3 \end{pmatrix}$

24. $\begin{pmatrix} 6 & 4 \\ 3 & 2 \end{pmatrix} - \begin{pmatrix} 7 & -3 \\ 1 & 4 \end{pmatrix}$

26. $\begin{pmatrix} -4 & -2 \\ 9 & 5 \end{pmatrix} - \begin{pmatrix} 6 & 3 \\ 4 & 6 \end{pmatrix}$

27. $\begin{pmatrix} -6 & 0 \\ 5 & 3 \end{pmatrix} - \begin{pmatrix} 3 & 4 \\ -1 & 2 \end{pmatrix}$

30. $\begin{pmatrix} 8 & 9 \\ 7 & 3 \end{pmatrix} - \begin{pmatrix} -4 & -2 \\ -2 & -1 \end{pmatrix}$

28. $\begin{pmatrix} -4 & 5 \\ -5 & 2 \end{pmatrix} - \begin{pmatrix} 8 & 2 \\ 1 & 0 \end{pmatrix}$

31. $\begin{pmatrix} -6 \\ 4 \end{pmatrix} - \begin{pmatrix} -4 \\ 6 \end{pmatrix}$

29. $(-4 \quad 9) - (-4 \quad 8)$

32. $(6 \quad 8) - (-1 \quad 3)$

MULTIPLES OF MATRICES WITH NEGATIVE ENTRIES

Remember that $3 \times (-2) = -6$ and $-3 \times 2 = -6$
but $(-3) \times (-2) = 6$
Remember also that $3(-2)$ means $3 \times (-2)$

EXERCISE 11h Find:

1. $5 \times (-3)$

2. -5×8

3. $(-5) \times (-3)$

4. $2(-4)$

5. $(-3) \times (-2)$

6. -2×6

7. $3 \times (-4)$

8. $3(-2)$

9. $6(-2)$

10. -1×4

11. $(-2) \times (-1)$

12. $4 \times (-1)$

> Find $5\begin{pmatrix} 6 & -2 \\ -4 & 3 \end{pmatrix}$
>
> $5\begin{pmatrix} 6 & -2 \\ -4 & 3 \end{pmatrix} = \begin{pmatrix} 30 & -10 \\ -20 & 15 \end{pmatrix}$

Find:

13. $5\begin{pmatrix} 6 & -1 \\ -3 & 2 \end{pmatrix}$ **16.** $-3\begin{pmatrix} 4 \\ 1 \end{pmatrix}$ **19.** $3(4 \quad -3)$

14. $4\begin{pmatrix} 1 & 2 \\ -3 & 0 \end{pmatrix}$ **17.** $7\begin{pmatrix} -1 & 4 \\ 3 & 1 \end{pmatrix}$ **20.** $-2\begin{pmatrix} -2 \\ 5 \end{pmatrix}$

15. $2\begin{pmatrix} -6 \\ -3 \end{pmatrix}$ **18.** $-3\begin{pmatrix} 6 & 1 \\ 3 & 2 \end{pmatrix}$ **21.** $-1(6 \quad 3 \quad -2)$

MIXED QUESTIONS

EXERCISE 11i   Find:

1. $\begin{pmatrix} 5 & -4 \\ 3 & 2 \end{pmatrix} + \begin{pmatrix} -3 & 1 \\ -1 & 3 \end{pmatrix}$ **6.** $\begin{pmatrix} 4 & -2 \\ 4 & -5 \end{pmatrix} + \begin{pmatrix} -5 & 8 \\ 3 & -2 \end{pmatrix}$

2. $\begin{pmatrix} 5 & -4 \\ 3 & 2 \end{pmatrix} - \begin{pmatrix} -3 & 1 \\ -1 & 3 \end{pmatrix}$ **7.** $\begin{pmatrix} 4 & -2 \\ 4 & -5 \end{pmatrix} - \begin{pmatrix} -5 & 8 \\ 3 & -2 \end{pmatrix}$

3. $3\begin{pmatrix} 5 & -4 \\ 3 & 2 \end{pmatrix}$ **8.** $-2\begin{pmatrix} -5 & 8 \\ 3 & -2 \end{pmatrix}$

4. $(4 \quad -5) + (3 \quad -1)$ **9.** $(6 \quad -5 \quad 0) - (2 \quad -4 \quad 0)$

5. $6(4 \quad -4)$ **10.** $7(-3 \quad -1)$

11. $\begin{pmatrix} 0 & 7 \\ 4 & 6 \end{pmatrix} - \begin{pmatrix} 9 & 3 \\ 3 & 6 \end{pmatrix}$ **13.** $(-4 \quad 2 \quad 6) - (-8 \quad 2 \quad -3)$

12. $\begin{pmatrix} 5 & 2 \\ -3 & 4 \end{pmatrix} + \begin{pmatrix} 2 & 6 \\ 5 & -8 \end{pmatrix}$ **14.** $-5\begin{pmatrix} -1 & 3 \\ 0 & -4 \end{pmatrix}$

15. $3(-2 \quad 3 \quad 4)$

16. $\begin{pmatrix} 6 \\ 4 \end{pmatrix} + \begin{pmatrix} -6 \\ 1 \end{pmatrix}$

17. $4\begin{pmatrix} 2 & 0 \\ -3 & 5 \end{pmatrix}$

18. $(5 \quad -7) + (-3 \quad 9)$

19. $\begin{pmatrix} 4 & -9 \\ 2 & -6 \end{pmatrix} + \begin{pmatrix} 3 & -1 \\ -5 & 4 \end{pmatrix}$

20. $-2(5 \quad -2 \quad -6)$

In questions 21 to 23, several alternative answers are given. Write down the letter that corresponds to the correct answer.

21. $\begin{pmatrix} 5 & -1 \\ 4 & -3 \end{pmatrix} + \begin{pmatrix} 3 & -4 \\ 1 & 2 \end{pmatrix}$ is equal to

A $\begin{pmatrix} 8 & -3 \\ 5 & -1 \end{pmatrix}$

B $\begin{pmatrix} 2 & -5 \\ 3 & -5 \end{pmatrix}$

C $\begin{pmatrix} 4 & -1 \\ 5 & -1 \end{pmatrix}$

D $\begin{pmatrix} 8 & -5 \\ 5 & -1 \end{pmatrix}$

22. $\begin{pmatrix} -4 & -5 \\ 9 & 3 \end{pmatrix} + \begin{pmatrix} -3 & 5 \\ 1 & -2 \end{pmatrix}$ is equal to

A $\begin{pmatrix} -7 & 0 \\ 10 & 1 \end{pmatrix}$

B $\begin{pmatrix} -1 & -10 \\ 8 & 5 \end{pmatrix}$

C $\begin{pmatrix} -7 & -10 \\ 8 & 1 \end{pmatrix}$

D None of these

23. $-2\begin{pmatrix} 6 & -3 \\ -2 & 0 \end{pmatrix}$ is equal to

A $\begin{pmatrix} -12 & 6 \\ 4 & 0 \end{pmatrix}$

B $\begin{pmatrix} -12 & -6 \\ 4 & 0 \end{pmatrix}$

C $\begin{pmatrix} 12 & 6 \\ -4 & -2 \end{pmatrix}$

D $\begin{pmatrix} -12 & 6 \\ 4 & -2 \end{pmatrix}$

MIXED EXERCISES

EXERCISE 11j There are no negative numbers in this exercise.

1. Give the sizes of the matrices $\begin{pmatrix} 7 & 1 & 2 \\ 3 & 8 & 1 \end{pmatrix}$ and $\begin{pmatrix} 1 \\ 2 \\ 3 \end{pmatrix}$

2. Give the entry in a) the third row of the second column
 b) the second row of the third column

    of the matrix $\begin{pmatrix} 6 & 8 & 1 \\ 2 & 4 & 3 \\ 6 & 1 & 4 \end{pmatrix}$

3. Find $\begin{pmatrix} 6 & 1 \\ 0 & 4 \end{pmatrix} + \begin{pmatrix} 1 & 4 \\ 3 & 2 \end{pmatrix}$

4. Find $5 \begin{pmatrix} 3 & 1 & 2 \\ 2 & 3 & 4 \end{pmatrix}$

5. Find $\begin{pmatrix} 7 & 8 \\ 3 & 2 \end{pmatrix} - \begin{pmatrix} 2 & 3 \\ 1 & 1 \end{pmatrix}$

6. Is it possible to add the matrices $\begin{pmatrix} 1 \\ 3 \end{pmatrix}$ and $(2 \quad 6)$?

EXERCISE 11k

1. Give the sizes of the matrices $\begin{pmatrix} 6 \\ 8 \end{pmatrix}$ and $(3 \quad 2 \quad 4 \quad 1)$

2. Give the entry in a) the first row of the second column
 b) the second row of the first column

    of the matrix $\begin{pmatrix} 4 & -3 & 2 & 1 \\ 5 & 6 & 0 & 3 \end{pmatrix}$

3. Find $\begin{pmatrix} 3 & -2 \\ 4 & 1 \end{pmatrix} + \begin{pmatrix} 6 & 4 \\ -5 & 2 \end{pmatrix}$

4. Find $\begin{pmatrix} 6 & 2 \\ -5 & 3 \end{pmatrix} - \begin{pmatrix} 2 & 1 \\ 3 & 4 \end{pmatrix}$

5. Find $3 \begin{pmatrix} 5 & -3 & 4 \\ 1 & 0 & 6 \end{pmatrix}$

EXERCISE 11I In each of the following questions, several alternative answers are given. Write down the letter that corresponds to the correct answer.

1. The sum of $\begin{pmatrix} 4 & 3 \\ 1 & -2 \end{pmatrix}$ and $\begin{pmatrix} 0 & 5 \\ 1 & 2 \end{pmatrix}$ is

   A $\begin{pmatrix} 3 & 26 \\ -2 & 1 \end{pmatrix}$ **B** $\begin{pmatrix} 4 & 8 \\ 2 & 0 \end{pmatrix}$ **C** $\begin{pmatrix} 4 & 3 & 0 & 5 \\ 1 & -2 & 1 & 2 \end{pmatrix}$ **D** $\begin{pmatrix} 12 \\ 2 \end{pmatrix}$

2. The entry in the second row of the first column of $\begin{pmatrix} 5 & -2 \\ 6 & 1 \end{pmatrix}$ is

   A 6   **B** -2 **C** 5   **D** 1

3. $\begin{pmatrix} 3 & 6 \\ 4 & 5 \end{pmatrix} - \begin{pmatrix} 1 & 7 \\ -3 & 2 \end{pmatrix}$ is equal to

   A $\begin{pmatrix} 2 & 1 \\ 1 & 3 \end{pmatrix}$ **B** $\begin{pmatrix} -15 & 39 \\ -11 & 38 \end{pmatrix}$ **C** $\begin{pmatrix} 2 & -1 \\ 7 & 3 \end{pmatrix}$ **D** $\begin{pmatrix} 19 \\ 77 \end{pmatrix}$

4. $6\begin{pmatrix} 1 & 2 \\ 3 & 0 \end{pmatrix}$ is equal to

   A $\begin{pmatrix} 6 & 12 \\ 18 & 0 \end{pmatrix}$ **B** $\begin{pmatrix} 6 & 12 \\ 18 & 6 \end{pmatrix}$ **C** $\begin{pmatrix} 7 & 8 \\ 9 & 6 \end{pmatrix}$ **D** $\begin{pmatrix} 6 & 12 \\ 24 & 6 \end{pmatrix}$

5. The size of the matrix $\begin{pmatrix} 3 & 4 & 2 \\ 4 & 5 & 1 \end{pmatrix}$ is

   A 2×2 **B** 3×2 **C** 2×3 **D** 3×3

12 AVERAGES. SPEED

Five children have the following sums of money: 5 p, 10 p, 2 p, 3 p and 20 p. If the money is collected then the total is 40 p. If it is now shared out equally amongst the five children, they will each receive 40 p ÷ 5 i.e. 8 p.

This amount, 8 p, is called the average (or mean) amount of money. Notice that none of the children actually has 8 p; they all have different amounts. If the total is shared out equally however each will have the average amount.

$$\text{Average} = \frac{\text{total amount}}{\text{number of items}}$$

EXERCISE 12a

Four children have the following pocket-money: 20 p, 25 p, 15 p and 40 p. What is the average amount of pocket-money?

The total amount = (20 + 25 + 15 + 40) p
$$= 100 \text{ p}$$
The average amount $= \dfrac{100 \text{ p}}{4}$
$$= 25 \text{ p}$$

1. Six children have the following amounts of pocket-money: 40 p, 10 p, 25 p, 30 p, 50 p and 85 p. What is the average amount of pocket-money?

2. Four people spend the following amounts on a meal: £2, £3.50, £4 and £2.50. What is the average amount paid for the meal?

3. Five girls spend the following amounts on crisps: 20 p, 18 p, 15 p, 25 p and 22 p. What is the average amount spent on crisps?

4. Ten children have the following sums of money: 5 p, 6 p, 2 p, 7 p, 15 p, 10 p, 8 p, 5 p, 9 p and 3 p. What is the average?

5. Three people contribute the following amounts to charity: 80 p, 50 p and 20 p. What is the average amount contributed?

6. Five cars are sold for the following amounts: £700, £2500, £1020, £1000 and £280. What is the average price of a car?

The average is not always an amount that could actually exist.
If the total is 42 p and this has to be shared amongst 5 people, then the average is $\frac{42 \text{ p}}{5}$ i.e. 8.4 p.

EXERCISE 12b

Find the average of 32 p, 46 p, 28 p and 17 p

Total = (32 + 46 + 28 + 17) p
= 123 p

$$\text{Average} = \frac{\text{total}}{\text{number of items}}$$

$$= \frac{123 \text{ p}}{4}$$

$$= 30.5 \text{ p}$$

In each question from 1 to 5 find the average sum of money.

1. 6 p, 5 p, 4 p and 10 p.
2. £23, £40, £20, £15 and £2.10.
3. 45 p, 89 p, 34 p, 23 p and 6 p.
4. 5 p, 6 p, 7 p, 8 p, 9 p, 10 p, 11 p and 12 p.
5. 67 p, 23 p, 79 p and 94 p.

We can find the average of quantities other than money.
In each question from 6 to 12 find the average of the given quantities.

6. 9 m, 7 m, 3 m, 5 m, 6 m and 3 m.
7. 12 g, 6 g, 3 g, 5 g, 10 g and 9 g.
8. 14 cm, 21 cm, 16 cm and 19 cm.

Averages. Speed 155

9. 6 kg, 3 kg, 8 kg, 10 kg, 4 kg, 7 kg, 2 kg and 6 kg.

10. 20 km, 30 km, 35 km, 28 km and 20 km.

11. 4 mm, 5 mm, 3 mm, 6 mm, 3 mm, 6 mm, 7 mm, 3 mm, 8 mm and 10 mm.

12. 15 cm², 10 cm², 30 cm², 5 cm² and 8 cm².

13. Five families have the following numbers of children: 2, 1, 3, 2, 3. Find the average number of children in a family.

14. Four families have the following numbers of children: 2, 0, 3, 2. Find the average number of children in a family. (Notice that, although one family has no children, we still divide by 4 because there are 4 families.)

15. On Monday to Friday last week the numbers of cars left in a car park were 45, 52, 49, 51, 50. Find the average number of cars in the car park per day.

16. On seven consecutive days there were 6, 22, 23, 20, 18, 20 and 21 flights from a small airport. What was the average number of flights per day? Give your answer correct to one decimal place.

17. A traffic survey found that from Monday to Friday the cars passing along a road during the day numbered 100, 126, 95, 130 and 100. What was the average number of cars per day?

18. A sweet shop has eight jars of sweets on a shelf and they contain the following numbers of sweets: 120, 130, 132, 90, 125, 85, 12 and 102. What is the average number of sweets per jar?

19. From Monday to Saturday, a newsagent sold the following numbers of papers: 200, 250, 240, 230, 200 and 247. What was the average number of papers sold per day?

20. From Monday to Friday a gardener worked the following number of hours: $7\frac{1}{2}$, 8, $6\frac{1}{2}$, 7 and $6\frac{1}{2}$. On average, how many hours per day did he work?

21. In a group of seven pupils the average height of the four girls is 151 cm and the average height of the three boys is 158 cm.

Find a) the total height of the girls
b) the total height of the boys
c) the average height of the group.

In questions 22 to 25, several alternative answers are given. Write down the letter that corresponds to the correct answer.

22. The average of 6p, 8p, 4p and 10p is

  A 4p **B** 28p **C** 5p **D** 7p

23. The average of 5m, 7m, 9m and 15m is

  A 36m **B** 8m **C** 4m **D** 9m

24. The average of 7km, 9km, 10km and 6km is

  A 11km **B** 40km **C** 8km **D** 5km

25. The average amount collected from 5 people was 20p. Which of the following must be true?

  A each person gave 20p **B** the total collected was £1

  C the third person gave 20p **D** the last person gave more than anyone else.

SPEED

If a car is driven at a steady speed and covers 40 miles in 1 hour then we say its speed is 40 miles per hour or 40 mph.

EXERCISE 12c

> A car covers 55 miles in 1 hour. What is its speed?
>
> The car's speed is 55 mph.

Give the speeds in the following cases, assuming that the speeds are steady.

1. A man walks 4 miles in 1 hour.

2. A cyclist covers 10 miles in 1 hour.

3. A motorist covers 32 miles in 1 hour.

4. A plane flies 250 miles in 1 hour.

5. Concorde flies 1250 miles in 1 hour.

6. A motorcyclist covers 55 miles in 1 hour.

Averages. Speed   157

We can use other units. A speed of 10 kilometres per hour is written 10 km/h. A speed of 5 metres per second is written 5 m/s.

Give the speeds in the following cases.

7. A bird flies 2 m in 1 second.

8. A cyclist travels 6 m in 1 second.

9. A motorist drives 45 km in 1 hour.

10. A plane flies 1020 km in 1 hour.

11. A snail moves 2 m in 1 hour.

12. A cheetah runs 30 m in 1 second.

If a car covers 70 miles in 2 hours at a steady speed it covers 35 miles in 1 hour, so its speed is 35 mph.

To find the speed we divide the distance travelled by the time taken.

i.e.
$$\text{Speed} = \frac{\text{distance}}{\text{time}}$$

EXERCISE 12d

At a steady speed, a cyclist covers 60 km in 4 hours. What is her speed?

$$\text{Steady speed} = \frac{\text{distance}}{\text{time}}$$
$$= \frac{60}{4} \text{ km/h}$$
$$= 15 \text{ km/h}$$

Find the steady speeds in the following cases.

1. A man walks 12 km in 3 hours.

2. A cyclist travels 22 miles in 2 hours.

3. A motorist drives 126 miles in 3 hours.

4. A plane flies 1250 miles in 5 hours.

5. A motorcyclist travels 196 km in 2 hours.

6. A deer runs 36 m in 8 seconds.

In questions 7 to 10, several alternative answers are given. Write down the letter that corresponds to the correct answer.

7. At a steady speed, a walker covers 20 km in 4 hours. His speed is

    A 5 km/h   **B** 4 km/h   **C** 20 km/h   **D** $\frac{1}{5}$ km/h

8. At a steady speed, a motorboat travels 76 km in 4 hours. Its speed is

    A $\frac{1}{19}$ km/h   **B** 19 mph   **C** 76 km/h   **D** 19 km/h

9. At a steady speed, a marble rolls 15 m in 5 seconds. Its speed is

    A 15 m/s   **B** 5 m/s   **C** $\frac{1}{3}$ m/s   **D** 3 m/s

10. At a steady speed, a runner covers 120 m in 12 seconds. His speed is

    A 120 m/s   **B** 12 m/s   **C** 12 mph   **D** 10 m/s

Sometimes the time is not an exact number of hours or seconds.

At a steady speed, a cyclist travels 27 km in $2\frac{1}{4}$ hours. What is the speed?

$$\text{Steady speed} = \frac{\text{distance}}{\text{time}}$$

$$= \frac{27}{2\frac{1}{4}} \text{ km/h}$$

$$= \frac{27}{2.25} \text{ km/h}$$

$$= 12 \text{ km/h}$$

Find the speeds in the following cases, assuming that they are steady.

11. A motorist travels 63 km in $1\frac{1}{2}$ hours.

12. A walker covers 3 miles in $\frac{3}{4}$ hour.

13. A cyclist travels 21 km in 1 hour 45 minutes.

14. A plane flies 243 miles in $1\frac{1}{2}$ hours.

15. A train travels 200 miles in $2\frac{1}{2}$ hours.

16. A runner covers 100 m in $12\frac{1}{2}$ seconds.

17. A dog walks 21 m in $7\frac{1}{2}$ seconds.

AVERAGE SPEED

Suppose that we travel by bus, from one town to another 40 miles away, taking 2 hours.

At times the bus travels quickly, at other times slowly and sometimes, at traffic lights or at road junctions, it does not move at all.

The speed changes all the time; it is not a steady speed.

If we can imagine doing the same journey of 40 miles at a steady speed, and taking the same time, then the speed would be $\frac{40}{2}$ mph i.e. 20 mph.

This imaginary speed is called the average speed.

$$\text{Average speed} = \frac{\text{total distance}}{\text{total time}}$$

EXERCISE 12e

A car travels 60 km in a time of $1\frac{1}{2}$ hours. What is its average speed?

$$\text{Average speed} = \frac{\text{total distance}}{\text{total time}}$$
$$= \frac{60}{1\frac{1}{2}} \text{ km/h}$$
$$= \frac{60}{1.5} \text{ km/h}$$
$$= 40 \text{ km/h}$$

Find the average speeds in the following cases.

1. A cyclist travels 19 km in 2 hours.

2. A walker covers 9 miles in 3 hours.

3. A train travels 360 km in 4 hours.

4. A runner covers 1000 m in 200 s.

5. A motorist drives 48 miles in $1\frac{1}{2}$ hours.

6. A plane flies 1750 km in $2\frac{1}{2}$ hours.

7. A motorboat covers 15 km in $\frac{3}{4}$ hour.

8. A motorcyclist travels 90 km in $1\frac{1}{2}$ hours.

9. A marathon runner covers 26 miles in 4 hours.

10. A bullet moves 60 m in $\frac{1}{2}$ sec.

DISTANCE AND TIME

If a car travels at an average speed of 30 mph for 4 hours then it will travel 30×4 miles, i.e. 120 miles, in the 4 hours.

$$\text{Distance} = \text{average speed} \times \text{time}$$
$$\text{Time} = \frac{\text{distance}}{\text{average speed}}$$

EXERCISE 12f

A cyclist travels at an average speed of 10 mph for 2 hours. How far does he go?

Distance = average speed × time
= 10 × 2 miles
= 20 miles

Find the distances travelled in the following cases.

1. A car is driven at an average speed of 40 mph, for 3 hours.

2. A yacht sails at an average speed of 8 km/h for 5 hours.

3. A cyclist rides at an average speed of $7\frac{1}{2}$ mph for 2 hours.

4. A plane flies at an average speed of 1250 km/h for $5\frac{1}{2}$ hours.

5. A swimmer swims at an average speed of 2.2 m/s for 13.5 seconds.

6. An inter-city train is driven at an average speed of 125 mph for 1 hour 24 minutes.

7. A girl jogs for 20 minutes at an average speed of 12 km/h.

8. A motorcyclist rides for 12 minutes at an average speed of 70 mph.

> A plane flies 800 km at an average speed of 150 km/h. How long does it take?
>
> $$\text{Time} = \frac{\text{distance}}{\text{average speed}}$$
>
> $$= \frac{800}{150} \text{ hours}$$
>
> $$= 5\tfrac{1}{3} \text{ hours}$$
>
> $$= 5 \text{ hours } 20 \text{ minutes}$$

Find the times taken in the following cases. Give your answers in hours and minutes, or in seconds.

9. A car is driven 125 km at an average speed of 75 km/h.

10. A wheelbarrow is pushed along at an average speed of 0.6 m/s over a distance of 33 m.

11. A heavy load is being transported a distance of 147 km at an average speed of $10\tfrac{1}{2}$ km/h.

12. A bicycle is ridden at an average speed of 18 km/h over a distance of 45 km.

13. A glider travels 93 km at an average speed of 15 km/h.

14. Merton and Caxton are 63 km apart. Mr Abel takes $1\tfrac{3}{4}$ hours to drive from one to the other. What is his average speed?

15. During the night, a badger moves through a wood for 5 hours. Its average speed is $1\tfrac{1}{2}$ km/h. What distance does it cover?

16. In the eighteenth century, the mail coach from London to Bristol travelled the 119 miles at an average speed of 7 mph. How long did the journey take?

17. A hot-air balloon floats, carried along by the wind. It is in the air for $4\tfrac{1}{2}$ hours and travels at an average speed of $1\tfrac{1}{2}$ km/h. What distance does it travel?

18. The speed of sound is 330 m/s. How many seconds does the noise of an explosion take to travel 825 m?

MIXED EXERCISES

EXERCISE 12g

1. Find the average of 5, 19, 31, 22 and 16.

2. Find the average of £6, £11, £3 and £4.

3. A car travels 92 miles in 2 hours. What is its average speed?

4. A man walks for 3 hours at an average speed of 6 km/h. How far does he walk?

5. How long does a runner take to cover 100 m at an average speed of 8 m/s?

EXERCISE 12h

1. Find the average of 12, 7, 8, 13, 4, 3 and 9.

2. Find the average of 6.5, 3.2, 4.8 and 2.3.

3. The distance by rail from Paddington to Bristol Parkway station is 117 miles. If a train makes the journey in $1\frac{1}{2}$ hours, what is its average speed?

4. Jane takes her dog for a walk and they cover $3\frac{1}{2}$ km at an average speed of $2\frac{1}{2}$ km/h. How long does the walk last?

5. A lawn mower moves at 3 m/s for 6 seconds. How far does it move?

EXERCISE 12i

1. What is the average of 3, 14, 11, 2 and 5?

2. Find the average of 8.1 kg, 4.7 kg, 7.9 kg and 3.3 kg.

3. A tractor pulls a seed-drill across a field at an average speed of 3 m/s. If the time taken is 70 s, how wide is the field?

4. A girl takes 12 *minutes* to ride her horse along a country lane $2\frac{1}{2}$ miles long. At what average speed, in miles per hour, does her horse trot?

5. How long does a spider take to climb up a strand of web that is 1 m long, if the spider's average climbing speed is 1 *centimetre* per second?

Averages. Speed 163

EXERCISE 12j In each of the following questions, several alternative answers are given. Write down the letter that corresponds to the correct answer.

1. The average of 13, 7, 5 and 15 is

   A 40 **B** 4 **C** 10 **D** 6

2. A car travels 192 miles in 5 hours. Its average speed is

   A 3.84 mph **B** 192 mph **C** 960 mph **D** 38.4 mph

3. A train journey takes 4 hours at an average speed of 80 km/h. The distance travelled is

   A 320 km **B** 20 km **C** 80 km **D** 40 km

4. An arrow covers 82 m at an average speed of 100 m/s. The time taken is

   A 82 s **B** 100 s **C** 8.2 s **D** 0.82 s

13 POLYGONS AND TESSELLATIONS

POLYGONS

A polygon is a plane (flat) figure bounded by straight lines.

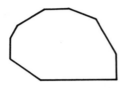

This is a nine-sided polygon.

Some polygons have names of their own:

a 3-sided polygon is a triangle

a 4-sided polygon is a quadrilateral

a 5-sided polygon is a pentagon

a 6-sided polygon is a hexagon

an 8-sided polygon is an octagon

REGULAR POLYGONS

A polygon is *regular* when *all its sides are the same length and all its angles are the same size.*

These are some regular polygons.

  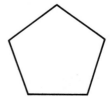

Polygons and Tessellations   165

EXERCISE 13a State whether or not each of the following figures is a regular polygon.

1. Rhombus
2. Square
3. Rectangle
4. Parallelogram
5. Isosceles triangle
6. Right-angled triangle
7. Equilateral triangle
8. Circle

REGULAR POLYGONS AND ROTATIONAL SYMMETRY

Regular polygons have rotational symmetry.

EXERCISE 13b Write down the order of rotational symmetry.

1.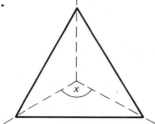

2.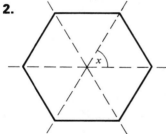

3.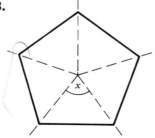

4.

5.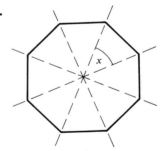

6.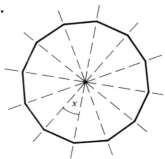

7. Find the size of the angle marked x for the polygons in questions 1 to 6.

8. Without drawing it, write down the order of rotational symmetry of a regular polygon with a) 7 sides b) 15 sides

DRAWING REGULAR POLYGONS

From the last exercise we can see that

> a regular n-sided polygon has rotational symmetry of order n.

We can use this property to make accurate drawings of regular polygons.

For example, we know that a regular 8-sided polygon (octagon) has rotational symmetry of order 8.

Therefore a rotation of $\frac{1}{8}$ of a revolution, leaves the polygon looking the same. ($\frac{1}{8}$ of a revolution is $\frac{1}{8}$ of 360° i.e. 45°)

To draw a regular octagon we start at the centre O and draw the 8 "spokes" of rotational symmetry so that each one is at 45° to the next.

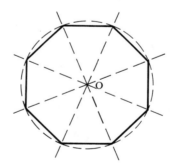

With centre O we draw a circle, then the points where the circle cuts the "spokes" are joined with straight lines. This gives the required octagon.

EXERCISE 13c Make an accurate drawing of each of the following polygons. Remember to use a *sharp* pencil. You will also need your protractor and a ruler.

1. A regular hexagon

2. An equilateral triangle

3. A square

4. A regular octagon

5. A regular pentagon

6. A regular 10-sided polygon

Polygons and Tessellations   167

THE EXTERIOR ANGLES OF A POLYGON

If we produce (extend) one side of a polygon, an angle is formed outside the polygon. It is called an exterior angle.

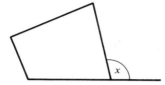

The angle marked x is an exterior angle.

To form all the exterior angles of a polygon we produce all the sides *in order*.

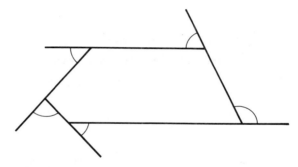

The diagram shows the five exterior angle of a pentagon.
The number of exterior angles is the same as the number of sides.

EXERCISE 13d **1.** In $\triangle ABC$ find
  a) the size of each marked angle
  b) the sum of the exterior angles

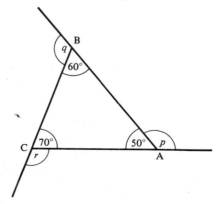

168   ST(P) Mathematics 3B

2. In the quadrilateral ABCD find
   a) the size of each marked angle
   b) the sum of the exterior angles.

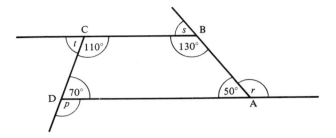

3. In the pentagon below find
   a) the size of each marked angle
   b) the sum of the exterior angles.

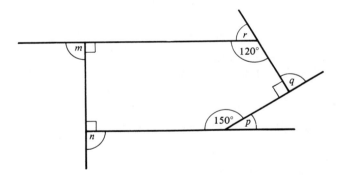

4. In △ABC find the sum of the exterior angles.

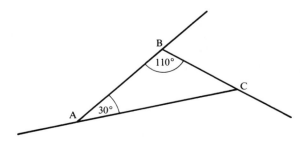

THE SUM OF THE EXTERIOR ANGLES OF A POLYGON

In the last exercise, the sum of the exterior angles came to 360° in each case. This is true of any polygon, whatever its shape or size.

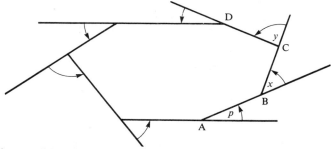

Imagine walking round this polygon. Start at A and walk along AB. When you get to B you have to turn through angle x to walk along BC. When you get to C you have to turn through angle y to walk along CD, ... and so on until you return to A. If you then turn through angle p you are facing in the direction AB again.

You have now turned through every exterior angle and have made one complete turn, i.e.

> the sum of the exterior angles of any polygon is 360°

EXERCISE 13e

Find the size of the angle marked x.

$$x + 100° + 60° + 90° + 50° = 360°$$
$$x + 300° = 360°$$
$$x = 60°$$

ST(P) Mathematics 3B

Find the size of the angle marked x.

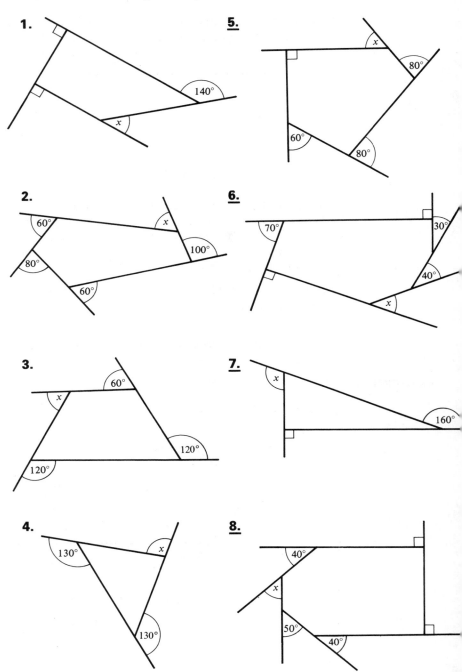

Polygons and Tessellations   171

9. Sketch a *regular hexagon*, and produce each side to form an exterior angle.

a) Write down the sum of the exterior angles.

b) Are the sizes of these angles related in any way?

c) Find the size of each exterior angle.

10. Repeat question 9 with a regular pentagon.

11. Repeat question 9 with a regular octagon.

12. Repeat question 9 with a regular 12-sided figure.

13.

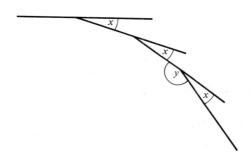

The diagram shows part of a regular 20-sided polygon.

Find a) the size of each exterior angle

b) the size of the angle marked y. (This is an interior angle.)

14. Use the ideas of question 13 to find the size of each interior angle in

a) a regular hexagon

b) a regular pentagon

15. In a regular polygon the size of each exterior angle is 20°. How many sides has the polygon?

16. In a regular 9-sided polygon find the size of each exterior angle.

INTERIOR ANGLES

The angles enclosed by the sides of a polygon are the interior angles. For example

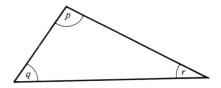

p, q and r are the interior angles of the triangle

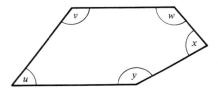

u, v, w, x and y are the interior angles of the polygon.

THE SUM OF INTERIOR ANGLES

We know that the sum of the exterior angles of any polygon is 360°.

We can use this fact to find the sum of the interior angles of any polygon.

Unless we are told that a polygon is regular we must assume that it is not. Consider a pentagon.

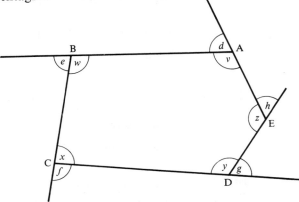

At each vertex the sum of the interior and exterior angles is 180°, e.g. at A, $v + d = 180°$

Polygons and Tessellations

There are five vertices, so the sum of *all* the exterior angles *and all* the interior angles is $5 \times 180° = 900°$

i.e. (sum of exterior angles) + (sum of interior angles) = 900°

But the sum of the exterior angles is 360°,

therefore   360° + (sum of interior angles) = 900°

Taking 360° from both sides gives

sum of interior angles = 900° − 360°

= 540°

This method works for any polygon: if a polygon has n sides, it has n vertices, therefore

(sum of exterior angles) + (sum of interior angles) = $n \times 180°$

i.e.   360° + (sum of interior angles) = $n \times 180°$

therefore sum of interior angles = $n \times 180° - 360°$

> The sum of the interior angles of an n-sided polygon is
> $180n° - 360°$

EXERCISE 13f

> Find the sum of the interior angles of a 9-sided polygon.
>
> Sum of interior angles = $n \times 180° - 360°$
>
> = $9 \times 180° - 360°$
>
> = $1620° - 360°$
>
> = $1260°$

Use the formula to find the sum of the interior angles in each case.

1. A pentagon

2. A 10-sided polygon

3. A hexagon

4. A quadrilateral

5. A 12-sided polygon

6. An octagon

For each of the following polygons find
a) the sum of all the interior angles   b) the size of the angle marked x.

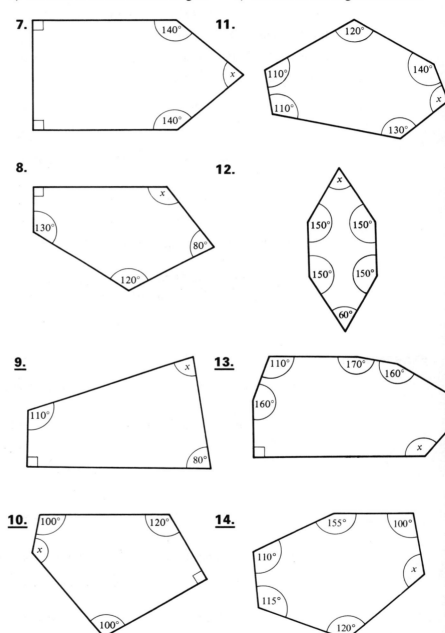

For a regular octagon find the size of
a) each exterior angle   b) each interior angle.

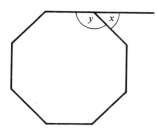

a) (The polygon is regular so the exterior angles are all the same size: there are 8 of them and they add up to 360°)

$$8x = 360°$$

$$x = 40°$$

Each exterior angle is 40°

b) (The polygon is regular so the interior angles are all the same size: at each vertex $x + y = 180°$)

$$40° + y = 180°$$

Take 40° from each side $y = 140°$

Each interior angle is 140°

For each of the following polygons find the size of
a) each exterior angle   b) each interior angle.

15. A regular pentagon

16. A regular hexagon

17. A regular octagon

18. A regular quadrilateral

19. A regular 10-sided polygon

20. A regular 12-sided polygon

21. A regular 20-sided polygon

22. A regular 3-sided polygon

23. In question 15 we found the size of an interior angle of a regular pentagon. Draw a rough sketch of the pentagon and mark in the size of each interior angle. Now draw, as accurately as you can, a regular pentagon with sides each 5 cm long.

24. Repeat question 23 for a regular hexagon.

TESSELLATIONS

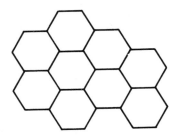

From the diagram we can see that regular hexagons fit together, without gaps, to form a flat surface.

When shapes do this we say that they *tessellate*.

Regular hexagons tessellate because each interior angle of a regular hexagon is 120°, so three vertices fit together to make 360°.

EXERCISE 13g

1. 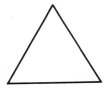 Trace this equilateral triangle and use it to cut out a template.

Using your template to draw round, show that you can cover an area of paper with equilateral triangles without any gaps, i.e. show that equilateral triangles tessellate.

2. Use squared paper to show that squares tessellate.

3. Regular hexagons, squares and equilateral triangles can be combined to make interesting patterns. Some examples are given below:

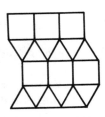 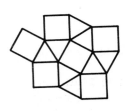 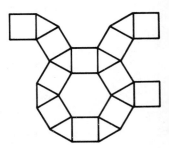

Copy these patterns and extend them.
(If you make templates to help you, make each shape of side 2 cm.)

Polygons and Tessellations 177

4. Make some patterns of your own using the shapes in question 3.

Not all regular polygons tessellate.

5.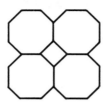

This is a pattern using regular octagons. They do not tessellate.

a) Explain why they do not tessellate.

b) What shape is left between the four octagons?

c) Continue the pattern. (Trace one of the shapes above, cut it out and use it as a template.)

6.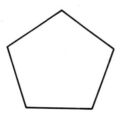

Trace this regular pentagon and use it to cut out a template.

a) Will pentagons tessellate?

b) Use your template to copy and continue this pattern until you have a complete circle of pentagons. What shape is left in the middle?

OTHER SHAPES THAT TESSELLATE

We can build up compound shapes using squares and equilateral triangles.

This shape, for example, is made from three squares and tessellates as shown below.

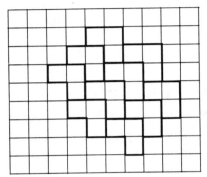 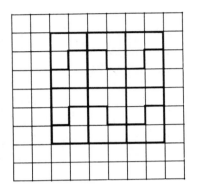

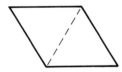

This shape is made from two equilateral triangles.

The diagram below shows a tessellation of the rhombus.

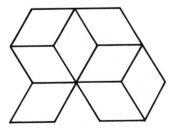

Not all compound shapes tessellate: this one, for example, does not.

Polygons and Tessellations 179

EXERCISE 13h Use squared paper to make some patterns of your own from tessellations of the following shapes.

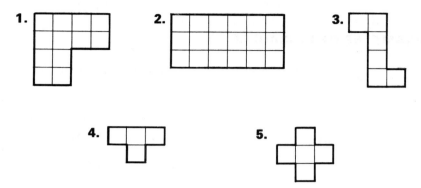

6. Make up your own shape from squares. Make a pattern from tessellations of your own shape but remember that not all shapes will work.

Trace the following shapes on to stiff paper. Cut them out and use them as templates to make tessellations.

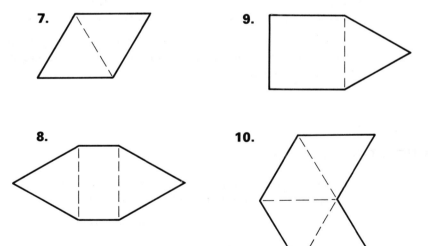

14 FORMULAE

EXAMPLES OF FORMULAE

The three sides of a triangle are of length a cm, b cm and c cm.

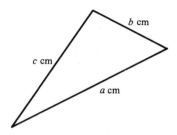

We find the perimeter by adding the three lengths together, so if the perimeter is p cm, then

$$p = a + b + c$$

This is a *formula* for the perimeter of a triangle when we are given the lengths of the three sides; p is the *subject* of the formula.

As a second example, we find the area of a rectangle by multiplying the length by the breadth. If the area is A cm², and the length and breadth are l cm and b cm, then

$$A = lb$$

USING A FORMULA

If we are given a formula $C = 2x + y$ and the values of x and y, we can use the formula to find the value of C

e.g., if $\qquad x = 3$ and $y = 4$

then $\qquad C = (2 \times 3) + 4$

$\qquad\qquad = 6 + 4$

$\qquad\qquad = 10$

We do not need to know what quantities the letters stand for to use the formula, but we do need to know the numbers.

EXERCISE 14a

> If $x = y + 3$, find x if $y = 6$
>
> $$y = 6$$
> $$x = y + 3$$
> $$= 6 + 3$$
> $$= 9$$

1. If $t = 3 + p$, find t if $p = 7$
2. If $a = b - 3$, find a if $b = 9$
3. If $g = h + 6$, find g if $h = 8$
4. If $x = 7 - y$, find x if $y = 3$

> If $C = a - p$, find C if $a = 10$ and $p = 6$
>
> $$a = 10 \text{ and } p = 6$$
> $$C = a - p$$
> $$= 10 - 6$$
> $$= 4$$

5. If $t = p + q$, find t if $p = 7$ and $q = 8$
6. If $a = b + c$, find a if $b = 10$ and $c = 9$
7. If $k = l - m$, find k if $l = 12$ and $m = 4$
8. If $s = t + u$, find s if
 a) $t = 3$ and $u = 13$
 b) $t = 31$ and $u = 46$
 c) $t = 3.4$ and $u = 4.8$

9. If $q = p - r$, find q if
 a) $p = 19$ and $r = 5$
 b) $p = 13.4$ and $r = 2.2$
 c) $p = 6.8$ and $r = 5.84$

> If $p = qr + 2$ find p if $q = 6$ and $r = 2$
>
> $q = 6, r = 2$
> $p = qr + 2$
> $\quad = (6 \times 2) + 2$
> $\quad = 12 + 2$
> $\quad = 14$

10. If $x = 4 + 3y$, find x if $y = 4$

11. If $a = 2b + c$, find a if $b = 5$ and $c = 6$

12. If $p = 5q - r$, find p if $q = 6$ and $r = 11$

13. If $a = bc + 7$, find a if $b = 4$ and $c = 2$

14. If $c = 6 + de$, find c if $d = 7$ and $e = 3$

15. If $x = yz - 2$, find x if $y = 3$ and $z = 4$

In questions 16 to 20, several alternative answers are given. Write down the letter that corresponds to the correct answer.

16. If $y = 5 + xz$, $x = 3$ and $z = 4$, then y is

    A 12   **B** 60   **C** 17   **D** 32

17. If $a = 6b + 7c$, $b = 3$ and $c = 2$, then a is

    A 13   **B** 33   **C** 18   **D** 32

18. If $p = qr - 3$, $q = 7$ and $r = 2$, then p is

    A 11   **B** 17   **C** 6   **D** 12

19. If $s = \frac{1}{2}(a + b + c)$, $a = 5$, $b = 7$ and $c = 6$ then s is

    A 18   **B** 9   **C** 36   **D** 105

20. If $d = ab + c$, $a = 3$, $b = 4$ and $c = 5$ then d is

    A 17   **B** 27   **C** 12   **D** 60

EXERCISE 14b

> If $s = t(u - v)$ find s if $t = 3$, $u = 6$ and $v = 2$
>
> $t = 3$, $u = 6$, $v = 2$
> $s = t(u - v)$
> $ = 3(6 - 2)$
> $ = 3 \times 4$
> $ = 12$

1. If $x = 2(4 - y)$, find x if $y = 1$
2. If $p = q(6 + r)$, find p if $q = 3$ and $r = 2$
3. If $a = 3(5 - b)$, find a if $b = 3$
4. If $g = h(l + k)$, find g if $h = 7$, $l = 4$ and $k = 3$
5. If $H = h(5 + j)$, find H if $h = 3$ and $j = 7$

> If $x = 4y^2$, find x if $y = 3$
>
> $y = 3$
> $x = 4y^2$
> $ = 4 \times y \times y$
> $ = 4 \times 3 \times 3$
> $ = 36$

6. If $p = q^2$, find p if $q = 6$
7. If $a = 6c^2$, find a if $c = 3$
8. If $x = 2 + y^2$, find x if $y = 4$
9. If $g = 5h^2$, find g if $h = 3$
10. If $x = yz^2$, find x if $y = 2$ and $z = 3$
11. If $y = x^2z$, find y if $x = 3$ and $z = 2$

> If $p = \dfrac{q}{r-s}$, find p if $q = 24$, $r = 10$ and $s = 6$
>
> $q = 24$, $r = 10$, $s = 6$
> $$p = \dfrac{q}{r-s}$$
> $$= \dfrac{24}{10-6}$$
> $$= \dfrac{24}{4}$$
> $$= 6$$

12. If $a = \dfrac{b+c}{3}$, find a if $b = 4$ and $c = 5$

13. If $p = \dfrac{4+q}{3+r}$, find p if $q = 8$ and $r = 1$

14. If $z = \dfrac{12-x}{y}$, find z if $x = 3$ and $y = 3$

15. If $p = \dfrac{2m}{n+3}$, find p if $m = 1$ and $n = 4$

16. If $x = 3y^2$, find x if $y = 7$

17. If $p = qr - 7$, find p if $q = 3$ and $r = 4$

18. If $g = \dfrac{h}{i+4}$, find g if $h = 6$ and $i = 8$

19. If $x = 6(y+z)$, find x if $y = 3$ and $z = 1$

In questions 20 to 22, several alternative answers are given. Write down the letter corresponding to the correct answer.

20. If $z = xy + 4$, $x = 6$ and $y = 7$, then z is

  A 38   **B** 46   **C** 17   **D** 66

21. If $a = \dfrac{b+c}{2}$, $b = 6$ and $c = 4$, then a is

  A 7   **B** 5   **C** 10   **D** $\dfrac{1}{5}$

22. If $p = q(4-r)$, $q = 6$ and $r = 2$, then p is

  A 22   **B** 36   **C** 8   **D** 12

USING NEGATIVE NUMBERS

EXERCISE 14c

> If $c = a - b$, find c if $a = -2$ and $b = -4$
>
> $a = -2$, $b = -4$
> $c = a - b$
> $ = (-2) - (-4)$
> $ = -2 + 4$
> $ = 2$

1. If $a = b + c$, find a if $b = -4$ and $c = -2$
2. If $a = 2b + c$, find a if $b = -3$ and $c = -4$
3. If $x = y - z$, find x if $y = -4$ and $z = -1$
4. If $s = t - u$, find s if $t = -2$ and $u = 4$
5. If $p = 2q + r$, find p if $q = 4$ and $r = -1$
6. If $a = b - 3c$, find a if $b = -6$ and $c = 2$
7. If $x = 3y - 2z$, find x if $y = -2$ and $z = 3$

> If $a = bc + 6$, find a if $b = -2$ and $c = -3$
>
> $b = -2$, $c = -3$
> $a = bc + 6$
> $ = (-2) \times (-3) + 6$
> $ = 6 + 6$
> $ = 12$

8. If $x = yz$, find x if $y = 2$ and $z = -4$
9. If $p = qr + 2$, find p if $q = -3$ and $r = 5$
10. If $a = 3 + bc$, find a if $b = 7$ and $c = -2$
11. If $y = xz$, find y if $x = -7$ and $z = -8$
12. If $d = ef + g$, find d if $e = -3$, $f = 4$ and $g = -2$

EXERCISE 14d

When a car travels for t hours at v mph, the distance, s miles, that it travels is given by the formula $s = vt$

If a particular car travels at 35 mph for 3 hours

a) what is the value of v?
b) what is the value of t?
c) use the formula to find the distance travelled.

a) $v = 35$

b) $t = 3$

c) $s = vt$
$ = 35 \times 3$
$ = 105$

The distance travelled is 105 miles.

1. The cost, C pence of x ices and y bars of chocolate is given by the formula

$$C = 20x + 15y$$

If 2 ices and 3 bars of chocolate are bought

a) what is the value of x?
b) what is the value of y?
c) use the formula to find the cost of 2 ices and 3 bars of chocolate.

2. The area, A m^2, of a rectangle measuring l m long by w m wide is given by the formula

$$A = lw$$

A rectangle is 6 m long and 3 m wide.

a) What is the value of l?
b) What is the value of w?
c) Use the formula to find the area of the rectangle.

3. A room is l m long, b m wide and h m high.
The area, A m², of the walls is given by the formula $A = 2h(l + b)$
A room is 6 m long, 3 m wide and 3 m high.

a) Give the values of l, b and h.

b) Use the formula to find the area of the walls of the room.

4. The length, l cm, of a rectangle whose area is A cm², and whose width is w cm, is given by the formula $l = \dfrac{A}{w}$
A rectangle is of area 36 cm² and its width is 3.6 cm.

Find its length in cm.

5. The cost, C pence, of x newspapers and y magazines is given by the formula $C = 40x + 70y$.
Use the formula to find the cost of 4 newspapers and 3 magazines.

Sometimes we are given the value of the subject of the formula and are asked to find the value of another letter.

We put the given numbers into the formula in their correct position and this gives an equation to be solved.

EXERCISE 14e

If $a = 2b + c$, find b when $a = 12$ and $c = 2$

$$a = 12, \; c = 2$$
$$a = 2b + c$$
$$12 = 2b + 2$$

Take 2 from each side $\quad 10 = 2b$

Divide each side by 2 $\quad 5 = b$

i.e. $\quad b = 5$

1. If $x = y + z$, find y when $x = 6$ and $z = 2$

2. If $p = q - r$, find q when $p = 3$ and $r = 4$

3. If $a = b + 2c$, find b when $a = 10$ and $c = 3$

4. If $x = 2y - z$, find z when $x = 12$ and $y = 8$

5. If $t = 3s + u$, find s when $t = 17$ and $u = 2$

> If $p = qr + 1$, find q if $p = 5$ and $r = 2$
>
> $\qquad p = 5, r = 2$
> $\qquad p = qr + 1$
> $\qquad 5 = q \times 2 + 1$
> i.e. $\qquad 5 = 2q + 1$
> Take 1 from each side $\qquad 4 = 2q$
> Divide each side by 2 $\qquad 2 = q$
> i.e. $\qquad q = 2$

6. If $x = yz$, find z when $x = 10$ and $y = 5$

7. If $s = tu$, find t when $s = 16$ and $u = 8$

8. If $p = tu + 4$, find u when $p = 16$ and $t = 3$

9. If $a = bc - 3$, find c when $a = 11$ and $b = 2$

10. If $d = 7 - ef$, find f when $d = 1$ and $e = 2$

> If $c = \dfrac{a}{b}$, find a if $c = 4$ and $b = 2$
>
> $\qquad c = 4, b = 2$
> $\qquad c = \dfrac{a}{b}$
> $\qquad 4 = \dfrac{a}{2}$
> Multiply each side by 2 $\qquad 2 \times 4 = \not{2} \times \dfrac{a}{\not{2}}$
> $\qquad 8 = a$
> i.e. $\qquad a = 8$

11. Given that $x = \dfrac{y}{z}$, find y if $x = 6$ and $z = 3$

12. Given that $p = \dfrac{q}{r}$, find q if $r = 3$ and $p = 7$

13. If $s = \dfrac{t}{u}$, find t if $s = 4$ and $u = 6$

14. If $l = \dfrac{m}{n}$, find m when $l = 2.5$ and $n = 3$

15. Given that $a = \dfrac{b}{c}$, find b when $c = 6$ and $a = 10$

16. Given that $r = \dfrac{p}{q}$, find p when $q = 3$ and $r = 12$

17. If $p = qr$, find q when $r = 7$ and $p = 21$

18. If $p = 2q + r$, find q when $r = 7$ and $p = 21$

19. If $p = q + 3r$, find q when $r = 7$ and $p = 21$

20. If $p = \dfrac{q}{r}$, find q when $r = 7$ and $p = 21$

USING NEGATIVE NUMBERS

EXERCISE 14f

If $r = s + t$, find t if $r = -8$ and $s = -2$

$$r = -8, \; s = -2$$
$$r = s + t$$
$$-8 = -2 + t$$

Add 2 to each side $\quad -8 + 2 = t$
$$-6 = t$$

i.e. $\quad t = -6$

1. If $x = y + z$, find y when $x = 4$ and $z = -2$

2. If $a = b + c$, find c when $a = -4$ and $b = 3$

3. If $p = q + r$, find q when $p = 6$ and $r = 9$

4. If $l = m + n$, find m when $n = -9$ and $l = -2$

5. If $s = t + u$, find u when $s = 7$ and $t = -2$

> If $f = g - h$, find h if $f = -6$ and $g = -4$
>
> $$f = -6, \ g = -4$$
> $$f = g - h$$
> $$-6 = -4 - h$$
> Add h to each side $\qquad -6 + h = -4$
> Add 6 to each side $\qquad h = -4 + 6$
> $$= 2$$

6. If $p = r - q$, find r when $q = 4$ and $p = -7$

7. If $s = t - u$, find u when $s = 7$ and $t = -2$

8. If $x = y - z$, find y when $x = -2$ and $z = -3$

9. Given that $x = y - z$, find z if $x = 6$ and $y = -2$

10. If $p = q - r$, find q when $p = 4$ and $r = -2$

> If $x = yz$, find z if $x = -12$ and $y = 3$
>
> $$x = -12, \ y = 3$$
> $$x = yz$$
> $$-12 = 3z$$
> Divide each side by 3 $\qquad -4 = z$
> i.e. $\qquad z = -4$

11. Given that $a = bc$, find c if $a = -12$ and $b = 3$

12. If $p = qr$, find q when $p = -12$ and $r = 6$

13. If $y = xz$, find z when $y = -7$ and $x = 2$

14. Given that $s = ut$, find u if $t = 8$ and $s = -32$

15. Given that $y = xz + 2$, find x if $y = -4$ and $z = 3$

> If $p = \dfrac{q}{r}$, find q if $p = -7$ and $r = 2$
>
> $$p = -7, \; r = 2$$
> $$p = \dfrac{q}{r}$$
> $$-7 = \dfrac{q}{2}$$
>
> Multiply each side by 2
> $$2 \times (-7) = \cancel{2} \times \dfrac{q}{\cancel{2}}$$
> $$-14 = q$$
>
> i.e. $\qquad q = -14$

16. Given that $z = \dfrac{x}{y}$, find x if $z = -6$ and $y = 7$

17. If $a = \dfrac{c}{b}$, find c if $a = -12$ and $b = 4$

18. If $p = \dfrac{r}{q}$, find r if $p = -2$ and $q = 10$

19. Given that $x = \dfrac{y}{z}$, find y if $x = -2$ and $z = 5$

20. If $s = \dfrac{t}{u}$, find t if $s = -9$ and $u = 4$

CHANGING THE SUBJECT OF THE FORMULA

A formula such as $a = b + c$ where a is the subject, can be rearranged to give a formula with b as the subject.

This is called *changing the subject*.

We deal with the formula in the same way as we would solve an equation.

EXERCISE 14g

> Given the formula $a = 2b + x$, make x the subject.
>
> $$a = 2b + x$$
>
> Take $2b$ from each side $\qquad a - 2b = x$
>
> i.e. $\qquad x = a - 2b$

1. Given the formula $p = x + q$, make x the subject
2. Given the formula $a = b + x$, make x the subject
3. Given the formula $c = x + d$, make x the subject
4. Given the formula $p = q + r$, make q the subject
5. Given the formula $x = y + z$, make y the subject

Given the formula $y = x - z$, make x the subject.

$$y = x - z$$
Add z to each side $\quad y + z = x$
i.e. $\quad x = y + z$

---

6. Given the formula $p = x - q$, make x the subject
7. Given the formula $b = x - c$, make x the subject
8. Given the formula $d = f - x$, make f the subject
9. Given the formula $g = h - f$, make h the subject

10. Given the formula $x = y - z$, make y the subject
11. Given the formula $a = b + c$, make b the subject
12. Given the formula $a = b - c$, make b the subject
13. Given the formula $m = n + p$, make p the subject

In questions 14 to 16, several alternative answers are given. Write down the letter that corresponds to the correct answer.

14. If $a = b + c$, then the formula with c as the subject is

    A $c = b - a$ **B** $c = a - b$ **C** $c = a + b$ **D** $c = \dfrac{a}{b}$

15. If $p = q + r$, then the formula with q as the subject is

    A $q = p + r$ **B** $q = r - p$ **C** $q = p - r$ **D** $q = pr$

16. If $x = y - z$, then the formula with y as the subject is

    A $y = x - z$ **B** $y = z + x$ **C** $y = z - x$ **D** $y = -x - z$

EXERCISE 14h

> If $a = bx$, make x the subject.
>
> $$a = bx$$
>
> (This is similar to $6 = 3x$; we would divide each side by 3)
>
> Divide each side by b $\qquad \dfrac{a}{b} = x$
>
> i.e. $\qquad x = \dfrac{a}{b}$

1. If $p = qx$, make x the subject

2. If $d = ex$, make x the subject

3. If $z = xy$, express x in terms of y and z (this means "make x the subject".)

4. Given the formula $t = us$, make u the subject

5. Given the formula $p = rq$, make r the subject

> If $b = \dfrac{x}{c}$, express x in terms of b and c
>
> $$b = \dfrac{x}{c}$$
>
> Multiply each side by c $\qquad c \times b = \overset{1}{\cancel{c}} \times \dfrac{x}{\cancel{c}_1}$
>
> $$bc = x$$
>
> i.e. $\qquad x = bc$

6. If $p = \dfrac{x}{n}$, express x in terms of p and n

7. Given that $t = \dfrac{x}{p}$, make x the subject of the formula

8. Given the formula $m = \dfrac{p}{n}$, make p the subject

9. If $z = \dfrac{y}{x}$, express y in terms of z and x

10. Given the formula $d = \dfrac{e}{f}$, make e the subject

MIXED OPERATIONS

EXERCISE 14i

1. If $x = 4 + y$, make y the subject of the formula.
2. Given the formula $p = q - r$, express q in terms of p and r.
3. If $w = uv$, make v the subject of the formula.
4. If $s = \dfrac{a}{b}$, make a the subject of the formula.
5. If $c = 2d + e$, make e the subject of the formula.
6. If $d = Hh$, make h the subject of the formula.
7. Given the formula $x = 9 + y$, express y in terms of x.
8. Given the formula $l = \dfrac{L}{k}$, express L in terms of l and k.
9. If $P + R = 4Q$, find R in terms of P and Q.

In questions 10 to 14, several alternative answers are given. Write down the letter that corresponds to the correct answer.

10. If $q = ut$, then the formula with t as the subject is

    A $t = \dfrac{u}{q}$ **B** $t = qu$ **C** $t = q - u$ **D** $t = \dfrac{q}{u}$

11. If $z = \dfrac{y}{x}$, then the formula with y as the subject is

    A $y = z + x$ **B** $y = zx$ **C** $y = \dfrac{z}{x}$ **D** $y = z - x$

12. If $w = uv$, then the formula with u as the subject is

    A $u = \dfrac{w}{v}$ **B** $u = w - v$ **C** $u = wv$ **D** $u = \dfrac{v}{w}$

13. If $h = \dfrac{k}{g}$, then the formula with k as the subject is

    A $k = \dfrac{h}{g}$ **B** $k = \dfrac{g}{h}$ **C** $k = h - g$ **D** $k = gh$

14. If $r = \dfrac{s}{t}$, then the formula with s as the subject is

    A $s = r - t$ **B** $s = t + r$ **C** $s = tr$ **D** $s = \dfrac{t}{r}$

MIXED EXERCISES

EXERCISE 14j
1. Given that $g = 2f + e$, find g if $f = 4$ and $e = 2$.
2. If $h = i + 2j$, express i in terms of h and j.
3. Given that $p = 3q - r$, find q if $p = 7$ and $r = 2$.
4. If $x = y - 2z$, make y the subject of the formula.
5. If $a = b(12 - c)$, find a if $b = 3$ and $c = 7$.

EXERCISE 14k
1. Given that $a + b - 2 = c$,
   a) find c when $a = 9$ and $b = 12$
   b) find a when $b = 6$ and $c = 10$
   c) find b when $a = 3$ and $c = 10$
   d) find c when $a = 11$ and $b = -3$
   e) express b in terms of a and c
   f) express a in terms of b and c.

2. Given that $A = mx + q$,
   a) find A if $m = 3$, $x = 11$ and $q = 4$
   b) find x if $A = 16$, $m = 3$ and $q = 4$
   c) find m if $x = 3$, $A = 12$ and $q = 6$
   d) find q if $A = 24$, $m = 1$ and $x = 1$
   e) find A if $m = 3$, $x = 4$ and $q = -5$
   f) express q in terms of A, m and x.

3. Using the formula $v = u + at$,
   a) calculate the value of v if $u = 6\frac{1}{2}$, $a = 4$ and $t = 3$
   b) find the value of a when $v = 50$, $u = 2$ and $t = 6$
   c) make u the subject of the formula.

4. The formula $F = \dfrac{9C}{5} + 32$ changes temperature from degrees Celsius to degrees Fahrenheit.
   a) Calculate F when $C = 25$
   b) Calculate F when $C = 35$

EXERCISE 14l In each question, several alternative answers are given. Write down the letter that corresponds to the correct answer.

1. If $x = y - z$, $y = 12$ and $z = 10$, then the value of x is

A 2   **B** 22   **C** 120   **D** none of these

2. If $p = 2(q + r)$, $q = 7$ and $r = 3$, then the value of p is

A 12   **B** 17   **C** 20   **D** 42

3. If $x = 2z + y$, then the formula with y as the subject is

A $y = \dfrac{x}{2z}$ **B** $y = 2z + x$ **C** $y = 2z - x$ **D** $y = x - 2z$

4. If $p = \dfrac{s}{t}$, then the formula with s as the subject is

A $s = p + t$ **B** $s = p - t$ **C** $s = pt$ **D** $s = \dfrac{p}{t}$

5. If $a = bx$, then the formula with b as the subject is

A $b = \dfrac{a}{x}$ **B** $b = \dfrac{x}{a}$ **C** $b = a - x$ **D** $b = ax$

EXERCISE 14m In each question, several alternative answers are given. Write down the letter that corresponds to the correct answer.

1. If $p = q + 2r$, $q = -2$ and $r = 3$, then the value of p is

A 8   **B** -8 **C** 4   **D** -4

2. If $x = 3(y - z)$, $y = 4$ and $z = -1$, then the value of x is

A 13   **B** 15   **C** 9   **D** 11

3. If $p = 2q + r$, $p = 9$ and $q = -3$, then the value of r is

A 3   **B** 15   **C** 12   **D** 24

4. If $a = 4b + c$, then the formula with c as subject is

A $c = \dfrac{a}{4b}$ **B** $c = a - 4b$ **C** $c = a + 4b$ **D** $c = 4ab$

5. If $z = 9x$, then the formula with x as the subject is

A $x = 9z$ **B** $x = z - 9$ **C** $x = z + 9$ **D** $x = \dfrac{z}{9}$

15 SIMPLE INTEREST

If you borrow money, you will normally have to pay for it, i.e. you must repay more than you borrow.

The difference between what you borrow and what you repay is called the *interest*. The sum of money borrowed (or lent) is called the *principal*. If the interest is an agreed percentage of this sum it is called *simple interest*.

In Book 2, we developed the formula

$$I = \frac{PRT}{100}$$

where I is the simple interest, P the principal, R the rate per cent each year (per annum or p.a.), and T the time in years for which the principal is borrowed or invested.

Hence $\quad\boxed{100\,I = PRT}$

This formula enables us to find any one of four quantities, if the other three are known.

EXERCISE 15a

Find the simple interest on £280 invested for 3 years at 7% p.a.

$$100\,I = PRT$$
$$P = 280,\ R = 7,\ T = 3$$
$$100\,I = 280 \times 7 \times 3$$
$$100\,I = 5880$$
$$100\,I \times \frac{1}{100} = 5880 \times \frac{1}{100}$$
$$I = 58.80$$

i.e. the simple interest is £58.80

Find the simple interest on:

1. £200 invested for 2 years at 8% p.a.
2. £160 invested for 5 years at 14% p.a.
3. £700 invested for 3 years at 13% p.a.
4. £330 invested for 7 years at 8% p.a.
5. £650 invested for 4 years at 6% p.a.
6. £420 invested for 3 years at 10% p.a.
7. £250 invested for 6 years at 12% p.a.
8. £850 invested for 5 years at 9% p.a.
9. £500 invested for 8 years at 12% p.a.
10. £720 invested for 3 years at 7% p.a.

What sum of money invested for 4 years at 8% p.a. gives £64 simple interest?

$$100 I = PRT$$
$$I = 64, \ R = 8, \ T = 4$$

therefore
$$100 \times 64 = P \times 8 \times 4$$
$$6400 = 32 \times P$$
$$\overset{200}{\cancel{6400}} \times \frac{1}{\cancel{32}} = \overset{1}{\cancel{32}} \times P \times \frac{1}{\cancel{32}}$$
$$200 = P$$

i.e. the principal is £200

---

11. What sum of money invested for 5 years at 12% p.a. gives £264 simple interest?

12. For how long must £370 be invested at simple interest of 9% p.a. to give interest of £233.10?

13. Find the annual rate per cent that earns £260 simple interest when £650 is invested for 5 years.

14. What sum of money earns £312 simple interest if invested for 8 years at 13% p.a.?

15. Find the annual rate per cent that earns £270 simple interest when £900 is invested for 6 years.

16. What sum of money invested for 6 years at 10% p.a. gives £270 simple interest?

17. For how long must £280 be invested at 5% p.a. to give simple interest of £70?

18. What rate per cent per annum gives simple interest of £150 on a principal of £1000 invested for 3 years?

19. What principal earns £352.80 simple interest if invested for 7 years at 9% p.a.?

20. What annual rate of simple interest is necessary to give interest of £416 on a principal of £800 invested for 8 years?

THE AMOUNT

The sum of the principal (P) and the interest (I) is called the amount.
This is denoted by A, i.e. $\quad A = P + I$

EXERCISE 15b

> What will £780 amount to if invested for 6 years at 11% simple interest per annum?
>
> The interest is found from $100 I = PRT$
>
> $$P = 780, \ R = 11, \ T = 6$$
>
> Therefore
> $$100 I = 780 \times 11 \times 6$$
> $$= 51\,480$$
> $$I = 514.80$$
>
> Amount = original sum + interest
> $$= £780 + £514.80$$
> $$= £1294.80$$

Find the amount if:

1. £120 is invested for 4 years at 15% p.a. simple interest.

2. £280 is invested for 5 years at 8% p.a. simple interest.

3. £530 is invested for 6 years at 9% p.a. simple interest.
4. £280 is invested for 2 years at 11% p.a. simple interest.
5. £680 is invested for 4 years at 7% p.a. simple interest.

What annual rate of simple interest enables £500 to amount to £900 in 10 years?

$$A = P + I \quad \text{i.e.} \quad I = A - P$$

Therefore $\quad I = 900 - 500 = 400$

but $\quad 100I = PRT$

where $\quad I = 400,\ P = 500,\ T = 10$

Therefore $\quad 100 \times 400 = 500 \times R \times 10$

$$R = \frac{100 \times 400}{500 \times 10} = 8$$

i.e. the rate of simple interest is 8% p.a.

What is the annual rate of simple interest if:

6. £500 amounts to £635 when invested for 3 years?
7. £380 amounts to £570 when invested for 5 years?
8. £150 amounts to £198 when invested for 4 years?

16 ANGLES IN A CIRCLE

THE BASIC FACTS

First we will revise some of the facts we already know about the circle.

1. Every point on a circle is the same distance from its centre. This distance is called the *radius* of the circle.

   Sometimes we use the word "circle" to include the space inside the curve. When we do this we call the curve itself the *circumference* of the circle.

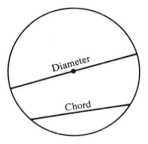

2. A straight line joining any two points on the circumference is called a *chord*.

3. Any chord passing through the centre of a circle is called a *diameter*.

We will now learn some new facts and definitions.

4.

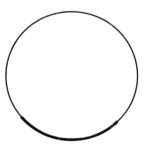

Any part of the circumference is called an arc. If the arc is less than half the circumference it is called a *minor arc*; if it is greater than half the circumference it is called a *major arc*.

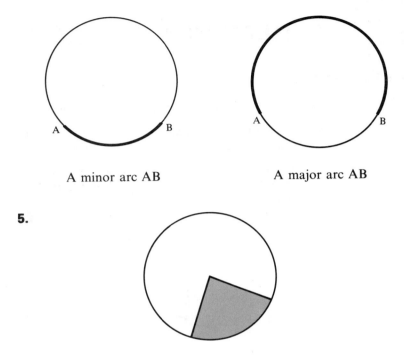

A minor arc AB   A major arc AB

5.

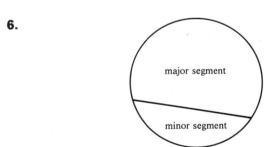

The shaded area is enclosed by two radii and an arc. It looks like a slice of cake and is called a *sector*.

6.

A chord divides a circle into two regions called segments. The larger region is called a *major segment* and the smaller region is called a *minor segment*.

Angles in a Circle   203

EXERCISE 16a **1.**

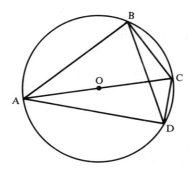

a) Name three chords that start at **A**.

b) Name three chords that start at **B**.

c) Name three chords that start at **C**.

d) Name three chords that start at **D**.

e) How many chords are there in this diagram?

f) Is any one of these chords a diameter? If so, name it.

Copy the following diagrams making them larger (a radius between 2.5 cm and 3 cm is suitable.) They do not need to be exact copies.

In questions 2, 3 and 4 draw all possible chords using only the letters already marked.

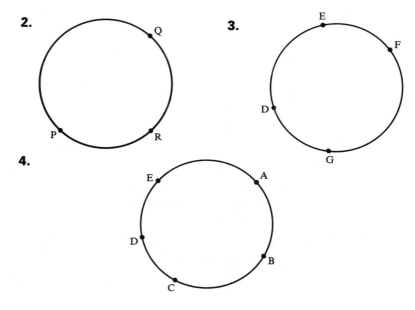

5.

Complete the following sentences.

a) The shaded area is a .... segment.

b) The unshaded area is a .... segment.

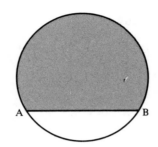

6.

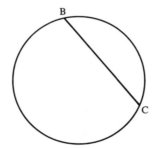

BC divides the circle into two segments.
Shade the minor segment.

7.

CD divides the circle into two segments.
Shade the minor segment.

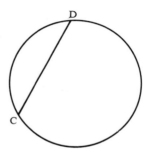

8.

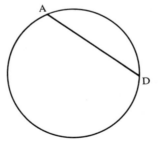

AD divides the circle into two segments.
Shade the major segment.

9.

Copy the given diagram twice.

a) Shade the major segment for the chord PR.

b) Shade the minor segment for the chord RS.

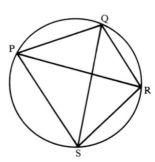

10.

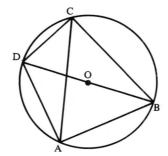

Copy the diagram three times.

a) Shade the minor segment for the chord BC.

b) Shade the major segment for the chord AC.

c) Shade the segment above the chord DB. How does the shaded area compare with the unshaded area? What do we call each of these areas?

11.

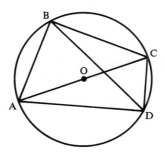

State whether the following statements are true or false.

a) The chord AD divides the circle into a major and a minor segment.

b) The chord BD divides the circle into two major segments.

c) The chord AC divides the circle into two equal segments.

d) The chord AC is a diameter.

e) The chord BD is a diameter.

12.

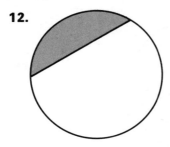

a) The shaded area is a major segment. (True or false?)

b) The unshaded area is a minor segment. (True or false?)

THE ANGLE SUBTENDED BY AN ARC

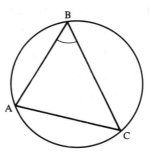

Consider a triangle ABC whose vertices A, B and C lie on a circle. The angle ABC is said to stand on the minor arc AC. We say that AC *subtends* the angle ABC at B, which is on the circumference.

Similarly the angle BAC stands on the arc BC, and BC subtends the angle BAC at A. Finally the angle ACB stands on the arc AB, and AB subtends the angle ACB at C.

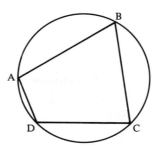

If the four vertices of a quadrilateral ABCD all lie on a circle we say that the quadrilateral is cyclic.

i.e. ABCD is a cyclic quadrilateral

EXERCISE 16b Questions 1 to 10 refer to the following diagram.

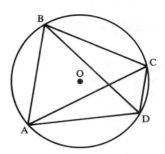

1. What arc does $A\hat{C}D$ stand on?
2. What arc does $A\hat{B}D$ stand on?
3. What arc does $B\hat{D}C$ stand on?
4. What arc does $B\hat{A}C$ stand on?
5. What arc subtends $B\hat{D}C$?
6. What arc subtends $B\hat{A}D$?
7. What arc subtends $D\hat{C}A$?
8. What arc subtends $C\hat{A}D$?
9. Name an angle at the circumference standing on
   a) the minor arc AC   b) the major arc AC.
10. Name an angle at the circumference standing on
    a) the minor arc BD   b) the major arc BD.

Questions 11 to 20 refer to the following diagram.

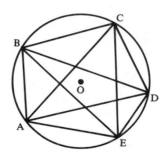

11. What arc does $B\hat{D}E$ stand on?
12. What arc does $C\hat{A}E$ stand on?
13. What arc does $D\hat{B}A$ stand on?
14. What arc does $D\hat{C}A$ stand on?
15. What arc subtends $E\hat{C}A$?
16. What arc subtends $B\hat{D}A$?
17. What arc subtends $A\hat{C}D$?
18. What arc subtends $D\hat{C}B$?
19. What angles stand on the minor arc AD?
20. What angle stands on the major arc BD?

DISCOVERING RELATIONSHIPS BETWEEN ANGLES

ANGLES IN THE SAME SEGMENT

EXERCISE 16c Copy the following diagrams making them at least twice as large. They need not be exact copies. For each diagram measure the angles marked by the letters.

1.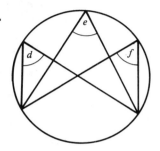

2.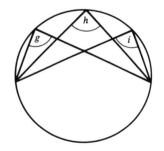

3.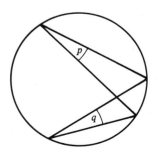

4.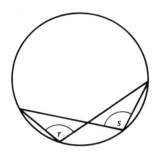

5.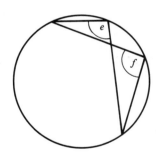

6.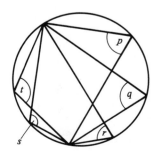

FIRST FACT

In the previous exercise accurate measurement should have shown that:

In question 1, $d = e = f$

In question 2, $g = h = i$

In question 3, $p = q$

In question 4, $r = s$

In question 5, $e = f$

In question 6, $p = q = r$ and $z = t$

These results give us our first important result for angles in a circle, namely that

> angles standing on the same arc of a circle and in the same segment are equal.

This is sometimes stated as

> angles in the same segment of a circle are equal.

EXERCISE 16d Use this result in questions 1 to 12 to find the angles marked by the letters.

1.

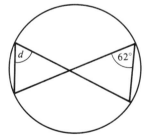

3.

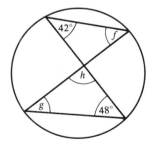

2.

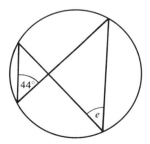

4.

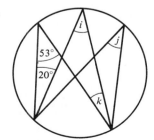

210 ST(P) Mathematics 3B

5.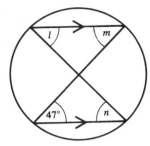

6.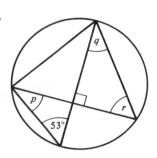

7.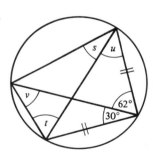

8.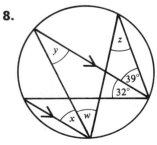

9.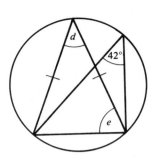

10.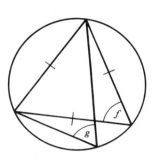

11.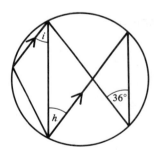

12.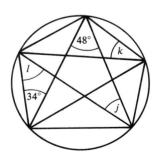

ANGLE AT THE CENTRE

EXERCISE 16e Copy the following diagrams making them at least twice as large. They need not be exact copies. For each diagram measure the angles marked by the letters.

1.
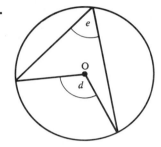

3.
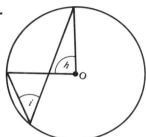

2.
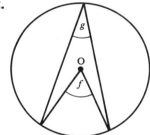

4.
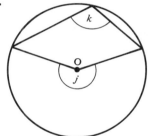

SECOND FACT

In the previous exercise, accurate measurement showed that:

In question 1, $d = 2e$

In question 2, $f = 2g$

In question 3, $h = 2i$

In question 4, $j = 2k$

These results give us our second important result for angles in a circle, namely that

> the angle which the arc of a circle subtends at the centre is equal to twice the angle it subtends at any point on the remaining circumference.

212   ST(P) Mathematics 3B

EXERCISE 16f Use this result in questions 1 to 8 to find the angles marked by the letters.

Reminder. Any triangle which has radii of a circle as two of its sides is an isosceles triangle. The base angles of an isosceles triangle are equal.

1.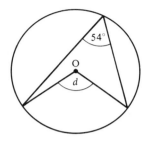

5.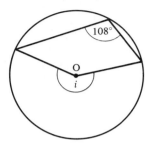

2.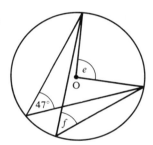

6.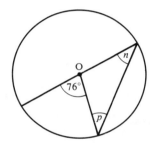

3.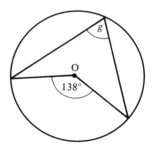

7.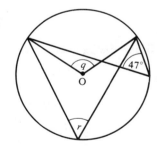

4.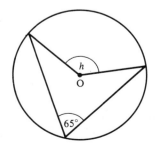

8.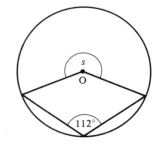

Angles in a Circle   213

9.

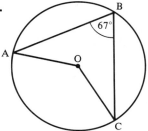

Calculate the size of AÔC.

10.

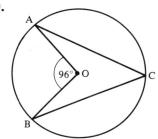

Calculate the size of AĈB.

11.

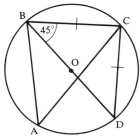

BC = CD. Find the size of
a) BD̂C   b) BÂC
c) BĈD

12.

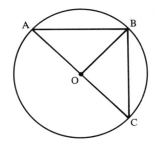

AC is a diameter.
BÔC = 90°. Find
a) OB̂C   b) OB̂A

13.

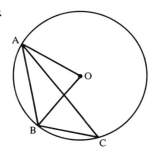

If AĈB = 48° find
a) AÔB   b) BÂO

14.

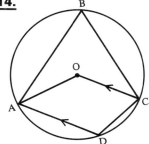

If AB̂C = 58° find
a) AÔC   b) OÂD

THIRD FACT

If we draw a diameter AB of a circle, it cuts the circumference into two semicircular arcs.

The angle subtended at the centre by a semicircle is 180°

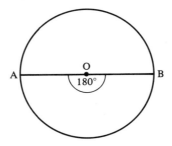

The angle subtended at a point on the circumference is half as big, i.e. 90°.

The angle in a semicircle is a right angle.

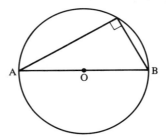

EXERCISE 16g Use this result in questions 1 to 4 to find the angles marked by the letters.

1.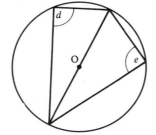

2.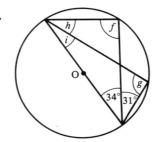

Angles in a Circle   215

3.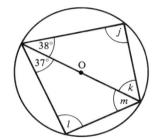

4.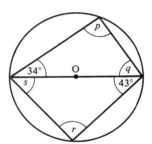

5.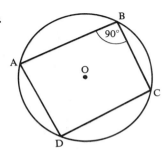

6.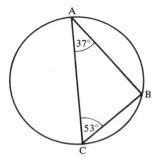

a) Does the chord AC pass through the centre of the circle O?

b) What is the value of ADC?

Does the chord AC pass through the centre of the circle?

MIXED EXERCISES

EXERCISE 16h In questions 1 to 10 find the angles marked by the letters.

1.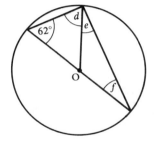

2.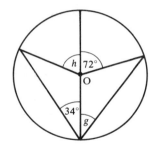

216 ST(P) Mathematics 3B

3.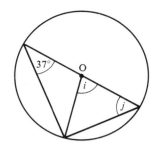

4.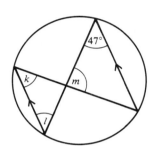

5.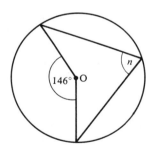

6.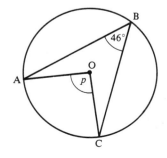

7.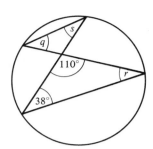

8.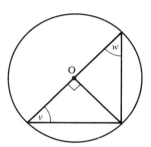

9.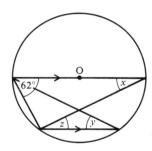

10.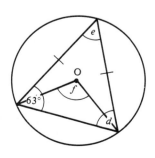

Angles in a Circle   217

EXERCISE 16i In this exercise several alternative answers are given. Write down the letter that corresponds to the correct answer.

Questions 1 and 2 refer to the diagram given below.

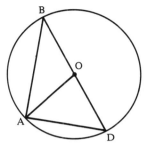

1. If BD is a diameter and $O\widehat{A}D = 50°$ the value of $B\widehat{A}O$ is

    A 50° **B** 90° **C** 100° **D** 40°

2. If $A\widehat{B}D = 40°$ the value of $A\widehat{O}D$ is

    A 50° **B** 40° **C** 20° **D** 80°

3.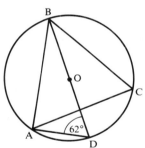

    If $B\widehat{D}A = 62°$ the value of $B\widehat{C}A$ is

    A 31° **B** 62° **C** 28° **D** 124°

4.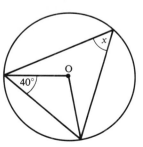

    The value of the angle marked x is

    A 100° **B** 40° **C** 50° **D** 80°

Questions 5 to 7 refer to the diagram given below.

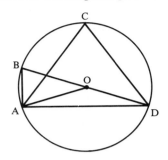

5. If $A\hat{C}D = 72°$ the value of $A\hat{O}D$ is

  A 144°   B 72°   C 18°   D 36°

6. If BD is a diameter the value of $B\hat{A}D$ is

  A 60°   B 180°   C 90°   D 45°

7. If $B\hat{D}C = 30°$ the value of $B\hat{A}C$ is

  A 60°   B 30°   C 90°   D 15°

8.

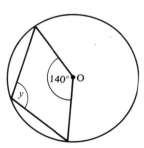

The value of the angle marked y is

  A 140°   B 110°   C 70°   D 90°

17 INEQUALITIES

SYMBOLS > AND <

In chapter 5 we introduced the idea of a directed number and the number line.

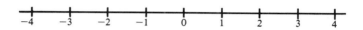

Every number, positive, negative or zero, has a position somewhere on the number line.

If a number a is to the right of a number b we say that
"a is greater than b" and we write $a > b$

e.g. $\quad\quad\quad 4 > 2, \quad 7 > 0, \quad 1 > -1, \quad -5 > -6$

On the other hand, if a number p is to the left of a number q on the number line, we say that "p is less than q" and we write $p < q$

e.g. $\quad\quad\quad 1 < 2, \quad -2 < 0, \quad -6 < -3, \quad -5 < 5$

EXERCISE 17a Insert < or > between each of the following pairs of numbers:

1.	9	4	**6.**	0	−2	**11.**	−5	12
2.	10	3	**7.**	−5	1	**12.**	−6	−4
3.	7	0	**8.**	−7	−2	**13.**	18	−3
4.	0	8	**9.**	8	−3	**14.**	−5	9
5.	5	7	**10.**	5	12	**15.**	−7	−16
16.	−4.1	−3.7	**20.**	−0.4	6.7	**24.**	−3.4	0.2
17.	−6.3	0.9	**21.**	−1.9	−2.8	**25.**	−7.3	−2.4
18.	0.2	−0.7	**22.**	−5.2	−4.7	**26.**	−3.7	−5.2
19.	−2.4	3.6	**23.**	9.2	10.6	**27.**	−4.2	3.1

THE MEANING OF AN INEQUALITY

Consider the statement $x > 3$

This is an *inequality*, whereas $x = 3$ is an equality or equation.

This inequality is true when x stands for any number that is greater than 3. There is a range of numbers that x can stand for and we can illustrate this range on a number line.

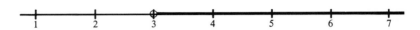

The circle at the left hand end of the range is "open" because x cannot be equal to 3 and therefore 3 is *not* included in the range.

EXERCISE 17b

Use a number line to illustrate the range of values of x for which a) $x < -1$ b) $x > -3$

a)

b)

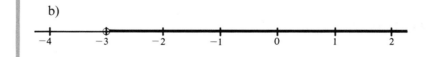

Use a number line to illustrate the range of values of x for which each of the given inequalities is true.

1. $x > 6$
2. $x < 5$
3. $x > -4$
4. $x < -2$

5. $x > 8$
6. $x < 3$
7. $x > -3$
8. $x < -1$

9. $x > 4$
10. $x < 6$
11. $x < -5$
12. $x > -7$

13. $x > 0$
14. $x > \frac{3}{4}$
15. $x < 0.75$

16. $x < 0$
17. $x > 0.25$
18. $x > -0.4$

19. $x < \frac{1}{2}$
20. $x < -0.5$
21. $x > -0.8$

> Does $x = -1$ satisfy the inequality $x > 4$
>
>
>
> Since -1 is to the left of 4 on the number line the inequality is not satisfied.

Decide whether or not the given value satisfies the given inequality.

22. $x = 3$, $x < 6$
23. $x = 5$, $x > -3$
24. $x = -4$, $x < 3$
25. $x = 0$, $x < -5$
26. $x = -2$, $x > -1$
27. $x = 7$, $x > 8$
28. $x = 8$, $x > 0$
29. $x = -3$, $x < 12$
30. $x = -2$, $x > -5$
31. $x = 0$, $x > -4$

> Give a whole number that satisfies the inequality $x < -3$
>
>
>
> x could be -4
>
> (Any number to the left of -3 on the number line would do, e.g. -5 or -100)

In the following questions give two whole numbers that satisfy the given equalities.

32. $x > 5$
33. $x > -3$
34. $x < -4$
35. $x > 5.7$
36. $x < -4.5$
37. $x > -0.2$
38. $x > 2$
39. $x > -7$
40. $x < -9$
41. $x < 3.2$
42. $x > -2.6$
43. $x < 0.9$

EXERCISE 17c

Give the smallest even whole number that satisfies the inequality $x > 10$

The next even number to the right of 10 on the number line is 12.

In questions 1 to 4, give the smallest whole number that satisfies the given inequality.

1. $x > 12$ **2.** $x > -4$ **3.** $x > 0$ **4.** $x > -18$

In questions 5 to 8, give the largest whole number that satisfies the given inequality.

5. $x < 8$ **6.** $x < 0$ **7.** $x < -5$ **8.** $x < -10$

In questions 9 to 12, give the smallest prime number that satisfies each of the inequalities.

9. $x > 8$ **10.** $x > 3$ **11.** $x > 9$ **12.** $x > 17$

In questions 13 to 16, give the largest prime number that satisfies each of the inequalities.

13. $x < 10$ **14.** $x < 8$ **15.** $x < 20$ **16.** $x < 3$

17. Give the largest whole number that is exactly divisible by 5 and which satisfies the inequality $x < 29$.

18. Give the smallest whole number that is exactly divisible by 7 and which satisfies the inequality $x > 36$.

19. Give the largest whole number that is exactly divisible by 2 and by 3 and which satisfies the inequality $x < 35$.

20. Give the smallest whole number that is exactly divisible by 3 and by 4 and which satisfies the inequality $x > 45$.

21. Give the largest whole number that is exactly divisible by 2, 3 and 5 and which satisfies the inequality $x < 50$.

ADDITION AND SUBTRACTION INVOLVING INEQUALITIES

Consider the inequality $4 > 2$

a) If we add 3 to each side we have $7 > 5$ and the inequality is still true.

b) If we subtract 1 from each side we have $3 > 1$, which is also true.

c) If we add -5 to each side we have $-1 > -3$, which is true.

d) If we subtract -3 from each side we have

$$4 - (-3) > 2 - (-3)$$

i.e. $$4 + 3 > 2 + 3$$

i.e. $$7 > 5 \text{ which is true.}$$

We conclude that

> an inequality remains true when the *same* number is added to, or subtracted from, both sides.

EXERCISE 17d

1. Subtract 3 from each side of the inequality $9 > 3$
2. Add 4 to each side of the inequality $7 < 12$
3. Subtract 6 from each side of the inequality $8 > -1$
4. Add 2 to each side of the inequality $-3 < 2$
5. Add -3 to each side of the inequality $5 < 9$
6. Subtract -5 from each side of the inequality $8 < 12$
7. Subtract -7 from each side of the inequality $9 < 16$
8. Add -4 to each side of the inequality $5 > -1$
9. Add -6 to each side of the inequality $-7 < -3$
10. Subtract 4 from each side of the inequality $-8 > -10$

Add 4 to each side of the inequality $x - 4 > 7$

$$x - 4 > 7$$

$$x > 11$$

11. Add 6 to each side of the inequality $x - 6 > 3$

12. Add 8 to each side of the inequality $x - 8 < 9$

13. Subtract 3 from each side of the inequality $x + 3 > 12$

14. Subtract 5 from each side of the inequality $x + 5 < 8$

15. Subtract 4 from each side of the inequality $x + 4 > 7$

16. Subtract 8 from each side of the inequality $x + 8 < 16$

17. Add -3 to each side of the inequality $x + 3 < -3$

18. Add -5 to each side of the inequality $x + 5 > -9$

19. Add 10 to each side of the inequality $x - 10 > -14$

20. Subtract 6 from each side of the inequality $x + 6 < -7$

SOLVING AN INEQUALITY

From the last exercise we can see that an inequality remains true when the *same* number is added to, or subtracted from, both sides.

Solving an inequality means finding the set of values of x for which the inequality is true.

EXERCISE 17e

Solve the inequality $x - 5 < 9$ and illustrate your answer on a number line.

$$x - 5 < 9$$

Add 5 to each side $\qquad x < 14$

Solve the following inequalities illustrating each answer on a number line.

1. $x - 3 < 6$ **6.** $x - 7 < 4$

2. $x - 4 > 5$ **7.** $x - 3 > 8$

3. $x + 3 > 7$ **8.** $x + 4 > 2$

4. $x + 5 < 8$ **9.** $x - 4 > 2$

5. $x + 3 > 2$ **10.** $x + 6 > 5$

Inequalities 225

11. $x - 4 > -1$
12. $x - 2 < -5$
13. $x + 2 < -3$
14. $x + 2 > -8$
15. $x - 9 > -6$
16. $x - 12 > -8$
17. $x - 5 < -2$
18. $x + 4 < -5$
19. $x + 7 > -4$
20. $x - 5 < -8$

Solve the inequality $5 - x > 3$ and illustrate your answer on a number line.

$$5 - x > 3$$

Add x to each side $\qquad 5 > 3 + x$

Subtract 3 from each side $\qquad 2 > x$

i.e. $\qquad x < 2$

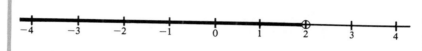

Solve the following inequalities, illustrating each answer on a number line.

21. $4 - x > 2$
22. $7 - x < 3$
23. $7 - x > -2$
24. $-3 - x > 2$

25. $-5 - x < 3$
26. $5 - x > 2$
27. $8 - x < 4$
28. $8 - x > -5$

29. $-4 - x > 5$
30. $-7 - x < 4$
31. $10 - x > 5$
32. $8 - x > -4$

33. $-7 - x < 2$
34. $9 - x < -5$
35. $8 - x > 8$
36. $4 - x > -3$

EXERCISE 17f In questions 1 to 6 multiply both sides of the inequality by the number in brackets. Is the inequality still true?

1. $9 > 7 \quad (2)$
2. $5 > -6 \quad (3)$
3. $2 < 4 \quad (5)$

4. $6 < 12 \quad (10)$
5. $35 > 20 \quad (3)$
6. $-10 < -6 \quad (4)$

In questions 7 to 12 multiply both sides of the inequality by the number in brackets. Is the inequality still true?

7. $12 > 5$ (-1)
8. $-7 < -4$ (-2)
9. $40 > 21$ (-3)
10. $3 < 8$ (-2)
11. $10 > -9$ (-5)
12. $-4 < 7$ (-4)

In questions 13 to 18 divide both sides of the inequality by the number in brackets. Is the inequality still true?

13. $4 > 2$ (2)
14. $8 < 12$ (4)
15. $-6 < 15$ (3)
16. $15 < 25$ (5)
17. $3 < 12$ (3)
18. $20 > -8$ (4)

In questions 19 to 24 divide both sides of the inequality by the number in brackets. Is the inequality still true?

19. $8 > 6$ (-2)
20. $25 < 50$ (-5)
21. $8 > -8$ (-4)
22. $12 < 36$ (-3)
23. $18 > 6$ (-6)
24. $-20 < 25$ (-5)

> An inequality remains true when both sides are multiplied or divided by the same positive number.

> Multiplication or division of an inequality by a negative number must be avoided because it makes the inequality untrue.

EXERCISE 17g

Solve the inequality $2x - 3 > 7$ illustrating the solution on a number line.

$$2x - 3 > 7$$

Add 3 to both sides $2x > 10$

Divide both sides by 2 $x > 5$

Inequalities 227

Solve the following inequalities illustrating each solution on a number line.

1. $3x - 2 < 7$
2. $1 + 2x > 3$
3. $4x - 3 > 9$
4. $3 + 5x < 8$
5. $5 + 2x < 6$
6. $3x + 1 > 8$
7. $4x - 5 < 4$
8. $6x + 5 > 18$
9. $9x - 4 > 14$
10. $3 + 4x < 11$
11. $7x - 3 < 14$
12. $4 - 9x > 1$

Solve the inequality $2x + 1 \leq 7 - 4x$ and illustrate the solution on a number line.

($\leq$ means "less than or equal to")
(As with equations we collect the letter term on the side with the greater number to start with. Remember, for instance, that $2x > -4x$)

$$2x + 1 \leq 7 - 4x$$

Add $4x$ to each side $\qquad 6x + 1 \leq 7$

Take 1 from each side $\qquad 6x \leq 6$

Divide both sides by 6 $\qquad x \leq 1$

(A solid circle is used for the end of the range because 1 *is* included)

Solve the following inequalities illustrating each solution on a number line.

13. $3 \leq 5 - 2x$
14. $4 - 3x > 1$
15. $2 \geq 13 - 5x$
16. $3 > 8 - 5x$
17. $4 - 7x \leq 11$
18. $10 - 3x > 2$
19. $7 - 2x \geq 6$
20. $5 > 4 - 5x$
21. $9 - 11x \leq 5$

22. $x + 3 > 5 - x$
23. $7x + 2 \leq 2x - 13$
24. $5x - 2 \leq 22 - 3x$
25. $4x - 3 > 9 - 2x$
26. $14 - 3x \leq 4 + 4x$
27. $x + 1 \leq 7 - 2x$
28. $10x - 3 \geq 3x - 17$
29. $11x + 10 > 2x - 26$
30. $5x - 7 \leq 3 - 2x$
31. $7x - 10 \leq 5 - 9x$

MIXED EXERCISE

EXERCISE 17h

1. Use a number line to illustrate the range of values of x for which each of the following inequalities is true.
 a) $x < -4$
 b) $x > -7$

2. Decide whether or not the given value satisfies the given inequality
 a) $x = -3$, $x > -4$
 b) $x = 7$, $x \leq 4$

3. Give two whole numbers that satisfy the inequality $x < -4.7$

4. Give the smallest whole number that satisfies the inequality
 a) $x > 7$
 b) $x > -5$

5. Give the largest whole number that satisfies the inequality
 a) $x < 12$
 b) $x < -5$

6. Solve the inequalities
 a) $x + 3 < -5$
 b) $9 - x > 4$
 c) $5x - 4 \geq 3 - 7x$

7. State whether the following statements are true or false
 a) $x = 3$ satisfies the inequality $x - 4 > 3$
 b) $x = -4$ satisfies the inequality $5x - 2 \leq 6$
 c) $x = 5$ does not satisfy the inequality $9x - 2 > 8$
 d) $x = -\frac{2}{3}$ does not satisfy the inequality $4x - 3 \leq 3 - 2x$

18 STATISTICS

BAR CHARTS AND FREQUENCY TABLES

A class was given a test which was marked out of 10. The marks are given below in the order in which the tests were checked.

$$8 \quad 6 \quad 10 \quad 10 \quad 8 \quad 6 \quad 7 \quad 9 \quad 5 \quad 5$$
$$7 \quad 7 \quad 5 \quad 4 \quad 3 \quad 7 \quad 5 \quad 8 \quad 10 \quad 9$$
$$9 \quad 7 \quad 8 \quad 7 \quad 6 \quad 3 \quad 8 \quad 6 \quad 7 \quad 7$$

When the numbers are written down in the order in which they arise they are called "raw data". *Data* means *pieces of information*.

This information needs sorting before it will tell us anything about how the class did in the test.

We find how many there are of each number, and form a *frequency table*.

Number of marks	3	4	5	6	7	8	9	10
Frequency (i.e. number of pupils)	2	1	4	4	8	5	3	3

We can now see, for example, that 7 is the mark which was gained by the greatest number of pupils.

This information can be illustrated by a bar chart.

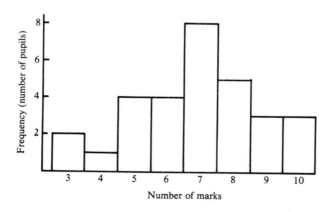

EXERCISE 18a **1.** This bar chart shows the shoe sizes of a class of girls.

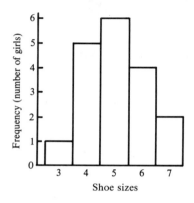

a) How many girls wear size 4 shoes?

b) How many girls wear size 7 shoes?

c) Copy and complete the following frequency table.

Shoe size	3	4	5	6	7
Frequency (i.e. number of girls)				4	

d) How many girls are there in the class?

2. This bar chart shows the different pets owned by pupils in a class.

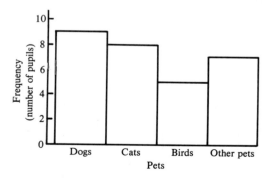

a) How many dogs are there?

b) How many birds are there?

c) Copy and complete the following frequency table.

Type of pet	Dog	Cat	Bird	Other pets
Frequency		8		

d) How many pets are there altogether?

e) Name some animals which would come under the heading "Other pets".

3. The pupils each recorded the number of pets owned by their family. This bar chart shows the information.

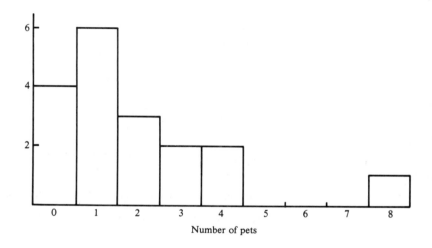

a) How many families own 2 pets each?

b) What is the most popular number of pets?

c) Copy and complete the following frequency table.

Number of pets	0	1	2	3	4	5	6	7	8
Frequency			3			0			

d) How many pets are there?

4. John counted the number of different vehicles passing the end of his road during a five minute interval. The information is given in the table.

Type of vehicle	Car	Lorry	Van	Motorcycle	Bicycle
Frequency	12	2	3	3	4

a) Draw a bar chart to illustrate this information. For the heights of the bars use 1 cm to represent 1 vehicle.

b) How many vehicles altogether passed the end of the road during the five minute interval?

c) What was the most common vehicle?

5. A choir is made up of men, women, boys and girls as shown in the table.

Men	Women	Boys	Girls
12	15	10	6

a) Draw a bar chart to show this information. For the height of the bars use 1 cm to represent 2 people.

b) Are there more males than females in the choir?

c) How many people are there in the choir?

6. A survey of cars in a local car park gave the following information about their country of origin.

Type of car	British	French	Japanese	American	Other
Frequency	31	16	28	4	21

a) Draw a bar chart to show this information. For the heights of the bars use 1 cm to represent 5 cars.

b) How many cars were there in the car park?

Statistics 233

MAKING A FREQUENCY TABLE

In the last exercise a bar chart or a frequency table was given with the numbers already sorted. If the information is given unsorted, like the array at the beginning of the chapter, then we need to make a frequency table from it.

A box contains 36 balloons. The balloons are taken out one at a time and their colours recorded. R stands for red, B for blue, Y for yellow, G for green and W for white.

R	B	B	Y	Y	G
G	W	G	G	W	R
R	W	R	B	R	B
B	R	G	R	G	W
G	W	Y	Y	B	R
Y	B	B	R	W	G

We find out how many there are of each colour by working down each column and putting a *tally mark*, /, against the colour in the table.

Do *not* go through the array counting how many Rs there are, then Bs and so on.

Colour	Tally	Frequency
R	~~////~~ ////	9
B	~~////~~ ///	8
Y	~~////~~	5
G	~~////~~ ///	8
W	~~////~~ /	6

Notice that every fifth tally mark is used to cross through the previous four to make a group of five. There are other ways of forming groups of five, e.g. ///// ///// or ⊠

EXERCISE 18b

1. In a class some pupils were wearing blazers, some pullovers, some cardigans and others none of these, over their shirts. The pupils each wrote on the board what they were wearing: B for blazer, C for cardigan, P for pullover and N for none of these. This was the result:

B	N	P	B
P	B	P	B
N	P	B	B
P	C	C	C
C	N	N	P
B	P	P	N

a) Copy and complete the table. (The tally marks for the first column have been filled in already. Check that you agree.)

Garment	Tally	Frequency
B	//	
C	/	
P	//	
N	/	

b) Draw a bar chart illustrating this information. For the height of the bars use 1 cm to represent one garment.

c) How many pupils are there in the class?

d) Which is the most popular garment?

e) Which is the least popular?

f) How many more pupils wore a blazer than none of the garments mentioned?

2. Three coins were tossed at the same time and the number of heads, (0, 1, 2 or 3), appearing each time was recorded. The results were as follows:

$$\begin{array}{cccccc} 2 & 3 & 1 & 2 & 0 & 1 \\ 2 & 0 & 1 & 2 & 3 & 2 \\ 1 & 3 & 1 & 3 & 1 & 1 \\ 1 & 2 & 2 & 1 & 1 & 2 \end{array}$$

a) Make a table with these headings.

Number of heads	Tally	Frequency

b) Draw a bar chart illustrating this information.

c) How many times was the set of three coins tossed?

d) Which number of heads appeared most often?

3. The passengers getting on to a bus during its journey were placed in one of four categories. The information is given below. M stands for man, W for woman, B for boy, G for girl.

$$\begin{array}{cccccccc} M & W & W & W & B & G & W & M \\ G & G & G & M & B & M & G & M \\ B & G & B & B & M & M & G & B \\ W & W & W & W & M & M & W & M \end{array}$$

a) Make a table similar to the table used in question 1.

b) Draw a bar chart illustrating this information.

c) How many males were there on the bus and how many females?

d) How many more girls than boys were there on the bus?

4. The pupils in a class counted the number of rooms, apart from bathrooms and kitchens, in which they and their families lived. The information is given below:

$$3\ 5\ 7\ 3\ 7\ 5\ 7\ 5\ 2\ 4$$
$$4\ 4\ 6\ 6\ 7\ 6\ 6\ 5\ 1\ 5$$
$$4\ 2\ 5\ 6\ 1\ 5\ 7\ 6\ 3\ 3$$

a) Make a frequency table.

b) Draw a bar chart illustrating this information.

c) What is the most common number of rooms?

d) How many rooms altogether do the pupils in this class have amongst them?

GROUPING THE INFORMATION

The heights in cms of 30 pupils are as follows:

164 166 154 154 156 155 161 157 156 154
166 149 166 153 166 164 150 159 159 147
161 165 153 160 156 164 163 148 163 157

There are 16 different numbers in this collection and if we made a frequency table it would have 20 entries, most of them with a frequency of 0 or 1 or 2. It is better in a case like this to group some of the heights together.

Heights in cms	Tally	Frequency
146 to 150	////	4
151 to 155	//// /	6
156 to 160	//// ///	8
161 to 165	//// ///	8
166 to 170	////	4

Now we will find it easy to draw a bar chart. We can already see that most of the heights come in the middle of the table.

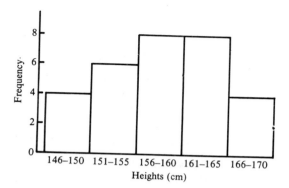

EXERCISE 18c **1.** The masses in kilograms of 30 pupils are as follows:

42 45 40 47 42 57
45 52 38 43 40 46
43 41 37 48 48 42
51 43 42 50 45 40
52 49 46 53 39 45

a) Copy and complete the following table.

Mass in kilograms	Tally	Frequency
35 to 39		
40 to 44		
45 to 49		
50 to 54		
55 to 59		

b) Draw a bar chart to illustrate this information. For the heights of the bars use 1 cm to represent 1 pupil.

2. The marks gained by pupils in an examination are given below.

$$\begin{array}{cccccccc}
40 & 84 & 49 & 52 & 75 & 31 & 56 & 59 \\
62 & 45 & 56 & 68 & 47 & 41 & 39 & 60 \\
70 & 51 & 63 & 38 & 85 & 35 & 53 & 86 \\
65 & 46 & 74 & 79 & 57 & 58 & 48 & 61 \\
\end{array}$$

a) Form a table as in question 1, using the groups 30 to 39, 40 to 49, 50 to 59, 60 to 69, 70 to 79 and 80 to 89.

b) Draw a bar chart to illustrate this information. For the heights of the bars use 1 cm to represent 1 pupil.

3. The times of arrival at school of 24 pupils were noted and the information is given below.

$$\begin{array}{cccccccc}
8.44 & 8.32 & 8.48 & 8.25 & 8.45 & 8.49 & 8.41 & 8.46 \\
8.43 & 8.35 & 8.46 & 8.26 & 8.46 & 8.42 & 8.35 & 8.31 \\
8.47 & 8.38 & 8.39 & 8.41 & 8.33 & 8.39 & 8.41 & 8.38 \\
\end{array}$$

a) Form a table as in question 1, using the groups 8.25 to 8.29, 8.30 to 8.34, 8.35 to 8.39, 8.40 to 8.44, and 8.45 to 8.49.

b) Draw a bar chart to illustrate this information. For the heights of the bars use 1 cm to represent 1 person.

4. The number of words in each of the sentences in the first page of a book were recorded. The information is given below.

$$\begin{array}{cccccccc}
9 & 26 & 11 & 15 & 21 & 19 & 29 & 19 \\
15 & 10 & 6 & 17 & 12 & 13 & 25 & 23 \\
11 & 4 & 13 & 25 & 21 & 17 & 16 & 13 \\
\end{array}$$

a) Form a table using the groups 1 to 5, 6 to 10, 11 to 15, 16 to 20, 21 to 25 and 26 to 29.

b) Draw a bar chart to illustrate this information.

PIE CHARTS

Statistics 239

We can use other types of diagram to illustrate the information we collect. One of these is a pie chart where the size of the slice is proportional to the frequency.

EXERCISE 18d

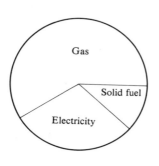

This pie chart shows the method of cooking in 60 houses in a street.

a) Which method is used least?

b) Which method is used most?

c) How does the number of houses using gas compare with the number using electricity?

a) Solid fuel is used least.

b) Gas is used most.

c) Gas is used in about twice as many houses as is electricity.

1.

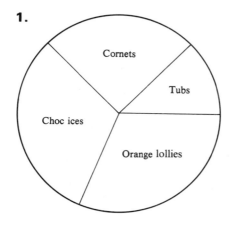

This pie chart shows the relative sales of different sorts of ice-cream.

a) Which is the least popular ice-cream?

b) The sales of one ice-cream form a quarter of the total sales. Which ice-cream is it?

c) Two ice-creams sell in equal quantities. Which two are they?

2.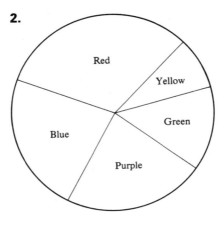

The pie chart shows the favourite colours of a group of people.

a) Which colour was the most popular?

b) Which colour was the least popular?

c) Which two colours were chosen by approximately the same number of people?

3.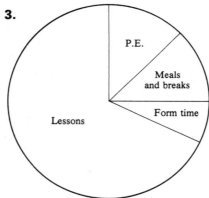

The pie chart shows the amounts of time spent on different activities in school.

a) Which activity has least time spent on it?

b) On which two activities are equal amounts of time spent?

4.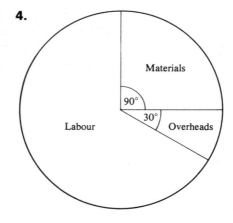

The pie chart shows the costs involved in making a television set. The total cost is £180.

a) What fraction of the total cost is the cost of materials?

b) What fraction of the total cost is the cost of overheads?

c) What is the cost of materials?

d) What is the cost of overheads?

e) What are the labour costs?

5.

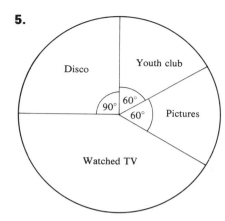

360 pupils were asked what they did on Saturday evening and the pie chart shows this information.

a) What fraction of the total number of pupils went to a disco?

b) What fraction of the total number of pupils went to a Youth Club?

c) How many went to a disco?

d) How many went to a Youth Club?

e) How many watched T.V.?

6.

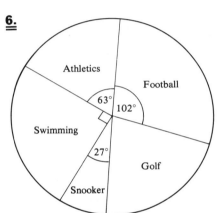

The pie chart shows the favourite sport of a group of 120 teenagers.

a) What fraction chose swimming?

b) How many chose athletics?

c) How many more chose football than snooker?

d) How many chose golf?

7.

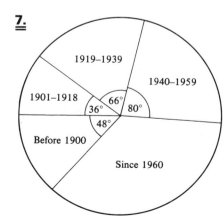

The pie chart shows the periods when the 180 houses in a village were built.

a) What fraction were built before 1940?

b) How many houses have been built since 1960?

c) How many more houses were built from 1940 to 1959 than from 1919 to 1939?

DRAWING PIE CHARTS

If we wish to draw pie charts we need to know the angle at the centre of the circle for each slice.

The table shows the number of items in a twenty-four piece tea set.

Pieces	Plates	Cups	Saucers	Other pieces
Number	7	6	6	5

The plates form $\frac{7}{24}$ of the number of pieces so the angle is $\frac{7}{24}$ of a complete turn, i.e. $\frac{7}{24} \times 360° = 105°$

The angle for the cups is $\frac{6}{24} \times 360° = 90°$

The angle for the saucers also is 90°

The angle for the other pieces is $\frac{5}{24} \times 360° = 75°$

To check the calculation we add the angles. The total should be 360°

Now draw a circle of radius 5 cm and draw one radius.

Use your protractor to draw slices of 105°, 90°, and 90°, turning the page to a suitable position to do so. The slice left should measure 75°. Label the pie chart as in previous examples.

EXERCISE 18e Draw pie charts to represent the following information. Work out the angles first.

1. In a class of 30, the eye colours of the pupils were recorded as follows.

Eye colour	Grey	Blue	Brown	Hazel
Frequency	8	4	14	4

2. In a class of 30, the means of transport for coming to school on a given day were recorded as follows.

Means of transport	Bus	Car	Bicycle	Walking	Other
Frequency	12	7	3	5	3

3. In a weekly timetable of 36 periods the distribution of time is as follows.

Subjects	Science, Maths	Art, Music	English	Languages	Others
Frequency	9	6	4	6	11

4. The times spent by the pupils in one form, watching different types of television programme one evening, were recorded.
What was the total viewing time?

Type of programme	Comedy Series	News	Plays and films	Documentaries	Other
Time (hours)	15	1	5	5	4

5. A group of 40 people were asked to name the sport they most enjoyed watching.

Sport	Soccer	Tennis	Cricket	Snooker	Other sports
Frequency	12	8	10	6	4

PICTOGRAPHS

If we are trying to attract people's attention to information, say on a poster, then a pictograph can be used.

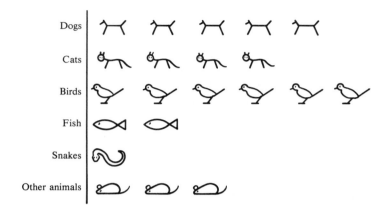

We can see at a glance without reading the names, that birds are the most popular pets amongst this group of families and snakes are the least popular.

Notice that all the drawings take up the same amount of room.

EXERCISE 18f

1. The pictograph shows the types of vehicle passing the school gate between 3.45 and 4.0 pm. one afternoon.

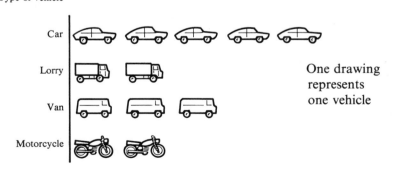

One drawing represents one vehicle

a) How many of each type of vehicle passed the gate?

b) How many vehicles passed altogether?

Statistics 245

2. The pictograph shows the number of people choosing the sport they most enjoyed playing.

Tennis	👤 👤 👤 大	👤 represents 5 people
Cricket	👤 👤 👤 👤 👤 ʃ	
Hockey	人	大 represents 4 people
Soccer	👤 👤 人	
Other	👤 大	人 represents 3 people and so on

a) How many people prefer to play tennis?

b) How many people prefer to play hockey?

c) How many people altogether chose a favourite sport?

3. When 25 people were asked in which country their car was made the following information was given.

Country of origin	Britain	U.S.A.	Japan	France	Other
Frequency	8	1	7	6	3

Draw a pictograph. Use the same simple symbol throughout and keep it the same size.

4. Draw a pictograph to illustrate the information given in Exercise 18e question 1. Use an eye ⦉●⦊ as the symbol.

5. Draw a pictograph to illustrate the information given in Exercise 18e question 2. The symbol to represent "other forms of transport" could be a horse or a plane.

19 CYCLIC QUADRILATERALS

REMINDER

In chapter 16 we learnt the following results:

1. All angles subtended at the circumference by the same arc are equal.

2. The angle which the arc of a circle subtends at the centre is equal to twice the angle subtended at the circumference by the same arc.

3. The angle in a semicircle is a right angle, i.e. it is 90°.

The first exercise in this chapter revises these results.

EXERCISE 19a In questions 1 to 6 find the angles marked by the letters.

1.
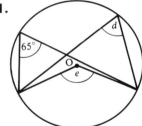

3.
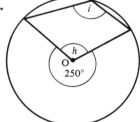

2.
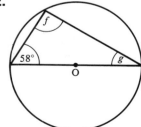

4.
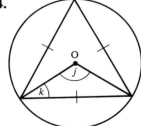

246

Cyclic Quadrilaterals 247

5.

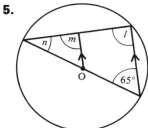

6.

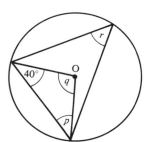

7.

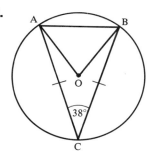

Find: a) AÔB
b) CÂB
c) OÂB

9.

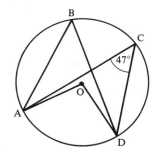

Find: a) AB̂D
b) AÔD

8.

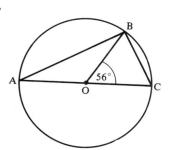

Find: a) OĈB
b) CÂB
c) OB̂A

10.

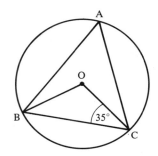

Find: a) OB̂C
b) BÔC
c) BÂC

THE OPPOSITE ANGLES OF A CYCLIC QUADRILATERAL

EXERCISE 19b Make drawings similar to the following diagrams making them at least twice as large. For each diagram measure the angles denoted by the letters. Can you see a relationship between the angles? (Be careful when you use your protractor because in each diagram at least one of the angles is obtuse.)

1.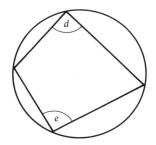

4.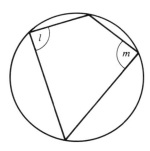

2.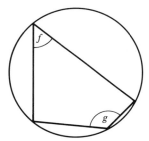

5.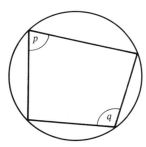

3.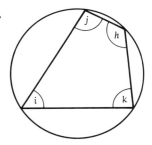

6.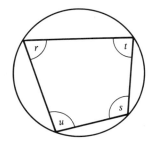

Cyclic Quadrilaterals

If you have measured the angles in exercise 19b accurately you will have found that:

In question 1 $d + e = 180°$

In question 2 $f + g = 180°$

In question 3 $h + i = 180°$ and $j + k = 180°$

In question 4 $l + m = 180°$

In question 5 $p + q = 180°$

In question 6 $r + s = 180°$ and $t + u = 180°$

FOURTH RESULT

These results give us the fourth important result about angles in a circle, i.e.

> the opposite angles of a cyclic quadrilateral add up to 180°   i.e. they are supplementary.

EXERCISE 19c Use this result to find the angles marked by the letters in questions 1 to 4.

1.
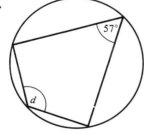

2.
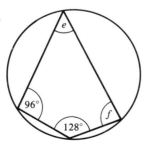

3.
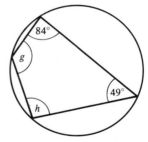

4.
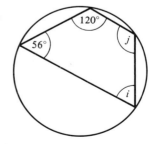

5.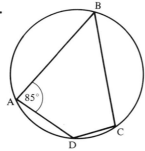

Find: BĈD

7.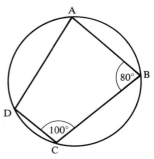

Find: a) BÂD
b) AD̂C

6.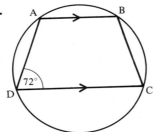

Find: a) AB̂C
b) DÂB
c) BĈD

8.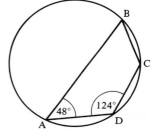

Find: a) AB̂C
b) BĈD

THE EXTERIOR ANGLE

EXERCISE 19d Make drawings similar to those given below making them at least twice as large. For each diagram measure the angles marked by the letters.

1.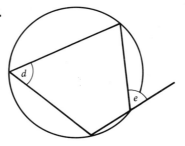

2.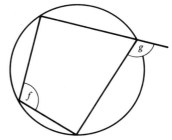

Cyclic Quadrilaterals 251

3.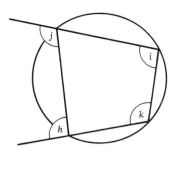

4.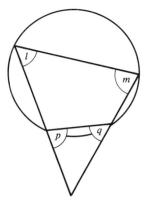

FIFTH RESULT

If you have measured the angles in questions 1 to 4 accurately you will have found that:

In question 1 $d = e$

In question 2 $f = g$

In question 3 $h = i$ and $j = k$

In question 4 $l = q$ and $m = p$

These results show that

> the exterior angle of a cyclic quadrilateral is equal to the interior opposite angle.

EXERCISE 19e In questions 1 to 4 find the angles marked by the letters.

1.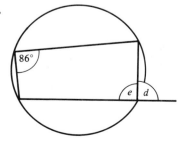

2.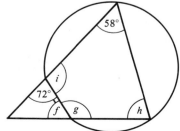

252 ST(P) Mathematics 3B

3.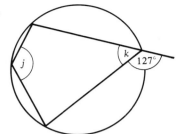

4.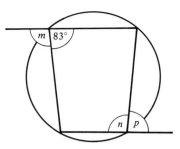

5.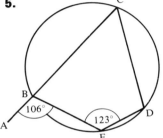

Find: a) $E\hat{D}C$ b) $B\hat{C}D$

6.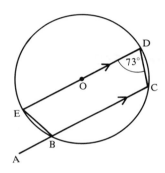

Find: a) $E\hat{B}A$ b) $B\hat{E}D$

MIXED EXERCISES

EXERCISE 19f This exercise uses all the circle results considered so far.

In questions 1 to 6 find the angles marked by the letters.

1.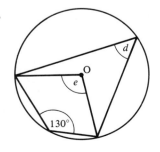

2.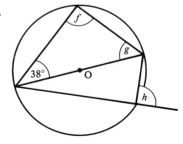

Cyclic Quadrilaterals   253

3.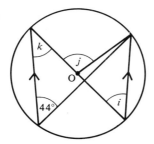

4.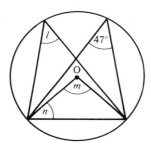

5.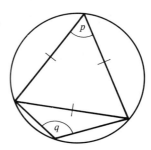

6.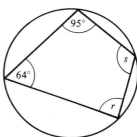

7.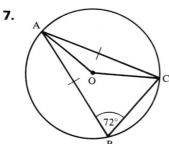

Find: a) AĈB   b) AÔC
      c) OĈA   d) OĈB

8.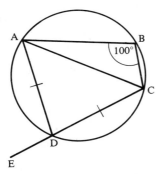

Find: a) AD̂C   b) AD̂E
      c) DÂC   d) AĈD

9.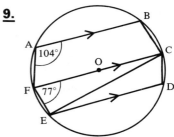

Find: a) BĈF   b) AB̂C
      c) CD̂E   d) FÊC

10.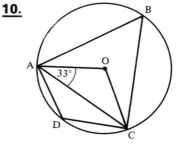

Find: a) OĈA   b) AÔC
      c) AB̂C   d) AD̂C

EXERCISE 19g In this exercise several alternative answers are given. Write down the letter that corresponds to the correct answer.

Questions 1 to 3 refer to the diagram given below.

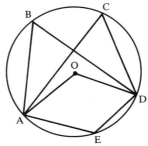

1. If $A\hat{O}D = 140°$ the value of $A\hat{C}D$ is

    A 280°   **B** 140°   **C** 70°   **D** 40°

2. If $A\hat{B}D = 70°$ the value of $A\hat{E}D$ is

    A 70°   **B** 110°   **C** 90°   **D** 140°

3. If $A\hat{B}D = 50°$ the value of $A\hat{O}D$ is

    A 100°   **B** 50°   **C** 130°   **D** 25°

Questions 4 to 6 refer to the following diagram.

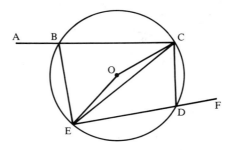

4. If $C\hat{D}F = 63°$ the value of $C\hat{B}E$ is

    A 126°   **B** 117°   **C** 27°   **D** 63°

5. If $A\hat{B}E = 108°$ the value of $E\hat{D}C$ is

    A 108°   **B** 142°   **C** 144°   **D** 136°

6. If $E\hat{D}C = 130°$ the value of $E\hat{O}C$ is

    A 50°   **B** 130°   **C** 25°   **D** 100°

20 RATIO AND PROPORTION

COMPARING SIZES OF QUANTITIES

If we have two quantities, such as 10 cm and 5 cm, we can compare their sizes in the following way,

10 cm compared with 5 cm is the same as 2 compared with 1.

If the two quantities are 12 g and 8 g then, because $12\,g = 3 \times 4\,g$ and $8\,g = 2 \times 4\,g$, we can say

12 g compared with 8 g is the same as 3 compared with 2.

EXERCISE 20a

> Compare 20 m with 16 m as simply as possible.
>
> $20\,m = 5 \times 4\,m$ and $16\,m = 4 \times 4\,m$
>
> so 20 m compared with 16 m is the same as 5 compared with 4.

Give the answers to the following questions as simply as possible.

1. Compare £10 with £2.
2. Compare 8 m with 6 m.
3. Compare 15 kg with 10 kg.
4. Compare 36 p with 12 p.
5. Compare 24 g with 18 g.

To save some of this writing we need a symbol that means "compared with". This symbol is :

"10 cm compared with 5 cm is the same as 2 compared with 1"
can be written   10 cm : 5 cm = 2 : 1

255

EXERCISE 20b Rewrite the following sentences using symbols.

1. 10 cm compared with 6 cm is the same as 5 compared with 3.
2. 15 p compared with 5 p is the same as 3 compared with 1.
3. £24 compared with £12 is the same as 2 compared with 1.
4. 72 cm compared with 27 cm is the same as 8 compared with 3.
5. 8 kg compared with 14 kg is the same as 4 compared with 7.
6. Rewrite your sentences from exercise 20a using symbols.

Write the following statements in words.

7. £18 : £12 = 3 : 2
8. 42 cm : 28 cm = 3 : 2
9. 16 cm : 24 cm = 2 : 3
10. 18 g : 20 g = 9 : 10

RATIO

We already know that 3 : 2 means "3 compared with 2".
Another way of expressing 3 : 2 is "the ratio 3 to 2"

i.e.   "the ratio 3 to 2"   means   "3 compared with 2"

SIMPLIFYING RATIOS

We know that 10 cm : 5 cm = 2 : 1. We can see that when both numbers of the ratio 10 : 5 are divided by 5 we get 2 : 1, so a ratio can be simplified if both parts are divided by the same number.

EXERCISE 20c

Simplify the ratio 20 : 15

20 : 15 = 4 : 3   (dividing each number by 5)

Simplify the ratio:

1. 8 : 10
2. 6 : 12
3. 16 : 8
4. 6 : 9
5. 70 : 50
6. 18 : 6
7. 24 : 20
8. 21 : 28
9. 27 : 18
10. 15 : 10
11. 12 : 9
12. 8 : 18

Ratio and Proportion

Sometimes two or more divisions are needed.

> Simplify the ratio 105 : 75
>
> 105 : 75 = 21 : 15   (dividing each number by 5)
> = 7 : 5   (dividing each number by 3)

Simplify the ratio:

13. 48 : 64   **17.** 64 : 144   **21.** 18 : 108
14. 72 : 168   **18.** 32 : 48   **22.** 175 : 200
15. 108 : 72   **19.** 75 : 120   **23.** 32 : 80
16. 270 : 150   **20.** 28 : 42   **24.** 288 : 120

> Simplify the ratio 18 m : 3 m
>
> (The units are the same for both parts)
> 18 m : 3 m = 18 : 3
> = 6 : 1

Simplify the ratio:

25. 16 cm : 48 cm   **27.** 44 g : 55 g   **29.** 72 kg : 36 kg
26. 20 p : 32 p   **28.** £25 : £15   **30.** 4 mm : 10 mm

EXERCISE 20d

> Simplify the ratio 180 cm : 2 m
>
> (The units must be the same so first change 2 m into cm.)
> 2 m = 2 × 100 cm = 200 cm
> 180 cm : 2 m = 180 cm : 200 cm
> = 18 : 20   (dividing by 10)
> = 9 : 10   (dividing by 2)

Simplify the following ratios. If the units are not the same, change the larger unit to the smaller unit.

1. 450 cm : 3 m

2. £2 : 75 p

3. 16 mm : 2 cm

4. £2.40 : 48 p

5. 1 m : 25 cm

6. 250 m : 0.5 km

7. 6 cm : 90 mm

8. £3.25 : 125 p

9. 90 p : £1.10

10. 25 m : 1000 cm

11. 35 cm : 5 mm

12. £14 : 70 p

In questions 13 to 16, several alternative answers are given. Write down the letter that corresponds to the correct answer.

13. 5 mm : 3 cm is the same as

A 5 : 3 **B** 1 : 6 **C** 50 : 3 **D** 1 : 15

14. 24 p : £1 is the same as

A 6 : 100 **B** 24 : 1 **C** 25 : 6 **D** 6 : 25

15. 30 cm : 2 m is the same as

A 30 : 2 **B** 3 : 2 **C** 3 : 20 **D** 3 : 200

16. 16 kg : 32 g is the same as

A 50 : 1 **B** 1 : 2000 **C** 1 : 2 **D** 500 : 1

SIMPLIFYING RATIOS BY MULTIPLYING

So far we have *divided* each part of a ratio by the same number to make the ratio simpler.

Sometimes we need to *multiply* each part of a ratio in order to simplify it,

e.g. if both parts of the ratio $\frac{1}{2}$: 4 are multiplied by 2, we get

$$\frac{1}{2} : 4 = 2 \times \frac{1}{2} : 2 \times 4$$
$$= 1 : 8$$

This is simpler because the ratio is expressed in whole numbers.

Ratio and Proportion 259

The next exercise gives ratios which can be simplified by multiplication.

EXERCISE 20e

> Simplify $\frac{1}{3} : 2$
>
> $\frac{1}{3} : 2 = 3 \times \frac{1}{3} : 3 \times 2$ (multiplying by 3)
>
> $= 1 : 6$

Simplify the following ratios.

1. $\frac{1}{2} : 2$
2. $\frac{1}{5} : 3$
3. $5 : \frac{2}{3}$
4. $3 : \frac{1}{4}$
5. $\frac{3}{4} : 2$
6. $6 : \frac{5}{7}$

> Simplify $6 : \frac{3}{4}$
>
> $6 : \frac{3}{4} = 4 \times 6 : 4 \times \frac{3}{4}$ (multiplying by 4)
>
> $= 24 : 3$
>
> $= 8 : 1$ (dividing by 3)

Simplify the following ratios.

7. $2 : \frac{2}{3}$
8. $\frac{3}{5} : 6$
9. $\frac{3}{4} : 3$
10. $\frac{4}{5} : 2$
11. $\frac{2}{3} : 4$
12. $\frac{5}{6} : 2$
13. $\frac{7}{3} : 14$
14. $\frac{1}{2} : \frac{1}{3}$
15. $\frac{2}{5} : \frac{1}{2}$

In questions 16 to 18, several alternative answers are given. Write down the letter that corresponds to the correct answer.

16. $\frac{5}{8} : 1$ is the same as

  A 5 : 1 **B** 8 : 5 **C** 5 : 8 **D** 1 : 8

17. $\frac{2}{3} : 2$ is the same as

  A 2 : 3 **B** 4 : 3 **C** 1 : 3 **D** 3 : 4

18. $4 : \frac{2}{5}$ is the same as

  A 4 : 2 **B** 10 : 1 **C** 2 : 5 **D** 5 : 8

PROBLEMS

EXERCISE 20f

> In a packet of 12 ball-point pens, there are 2 red ones and the rest are blue. What is the ratio of the number of blue pens to the number of red pens?
>
> There are 10 blue and 2 red pens
>
> Therefore the ratio of blue to red is $10:2$
>
> $= 5:1$

1. Peter is 16 years old and Anne is 12 years old. What is the ratio of Peter's age to Anne's age?

2. The pupils in a class own 16 pets between them. If 10 of the pets are dogs, what is the ratio of the number of dogs to the number of other animals?

3. Pat has 9 exercise books in her desk. If 3 are brown and the rest are green, what is the ratio of the number of brown books to the number of green books?

4. In a class of 24, there are 15 boys. What is the ratio of the number of girls to the number of boys?

5. Of 22 vehicles parked in a road, 6 are vans. What is the ratio of the number of vans to the number of other vehicles?

6. Driving the 120 miles from Abbotsford to Barford, I cover 96 miles on motorways. What is the ratio of the number of miles driven on motorways to the number of miles on other roads?

7. A farm of 200 hectares uses 125 hectares for growing cereals and the rest for grazing. What is the ratio of the number of hectares used for growing cereals to the number of hectares used for grazing?

8. On a day out Mr Jones spends £6.50 on tickets and £7.50 on food. What is the ratio of the amount spent on food to the amount spent on tickets?

Ratio and Proportion 261

We are not always asked to simplify a ratio by making both parts into whole numbers.

Sometimes we are asked to write the ratio so that one part is 1. This may mean that the other part is not a whole number.

For example, $\quad 2:3 = 1:1\frac{1}{2} \quad$ (dividing by 2)

and $\quad\quad\quad\quad 5:7 = 1:\frac{7}{5} \quad$ (dividing by 5)

$\quad\quad\quad\quad\quad\quad = 1:1\frac{2}{5}$ or $1:1.4$

We have written each ratio in the form $1:n$, where n may be a fraction or a decimal.

EXERCISE 20g

Complete the ratio $5:6 = 1:$

$5:6 = 1:\frac{6}{5} \quad$ (dividing each part by 5)

$\quad\quad = 1:1.2 \quad$ or $\quad 1:1\frac{1}{5}$

Complete the ratio

1. $4:7 = 1:$
2. $8:12 = 1:$
3. $5:3 = 1:$
4. $2:5 = 1:$
5. $10:7 = 1:$
6. $5:1 = 1:$

Write the ratio $6:15$ in the form $1:n$

$6:15 = 1:\frac{15}{6} \quad$ (dividing by 6)

$\quad\quad = 1:2.5 \quad$ or $\quad 1:2\frac{1}{2}$

Write the ratio in the form $1:n$

7. $5:7$
8. $10:9$
9. $2:3$
10. $8:3$
11. $4:9$
12. $5:2$

> Complete the ratio $4:5 = \quad :1$
>
> (Notice that this time it is the *second* part that is to be 1)
>
> $4:5 = \frac{4}{5}:1$ (dividing by 5)
>
> (This can be written as $0.8:1$)

Complete the following ratios.

13. $3:2 = \quad :1$ **15.** $12:4 = \quad :1$ **17.** $6:8 = \quad :1$

14. $6:5 = \quad :1$ **16.** $15:10 = \quad :1$ **18.** $20:4 = \quad :1$

Write the following ratios in the form $n:1$

19. $3:4$ **21.** $1:4$ **23.** $9:5$

20. $10:5$ **22.** $2:5$ **24.** $13:4$

In questions 25 to 27, several alternative answers are given. Write down the letter that corresponds to the correct answer.

25. The ratio $4:3$ is the same as

    A $1:3$ **B** $3:4$ **C** $1:\frac{4}{3}$ **D** $1:0.75$

26. The ratio $9:4$ is the same as

    A $2\frac{1}{4}:1$ **B** $1:2\frac{1}{4}$ **C** $4:9$ **D** $9:1$

27. The ratio $7:2$ is the same as

    A $1:3.5$ **B** $5:1$ **C** $3.5:1$ **D** $14:1$

COMPARING THREE QUANTITIES

We can compare three or more quantities by using ratio.

For example, $6\,\text{kg}:3\,\text{kg}:12\,\text{kg} = 6:3:12$

                                   $= 2:1:4$ (dividing by 3)

Ratio and Proportion 263

EXERCISE 20h

> Simplify the ratio $24 : 18 : 6$
>
> $24 : 18 : 6 = 4 : 3 : 1$ (dividing by 6)

Simplify the following ratios.

1. $9 : 18 : 6$
2. $6 : 8 : 10$
3. $8 : 12 : 16$
4. $21 : 14 : 42$
5. $12 : 36 : 24$
6. $22 : 11 : 33$
7. $72 : 54 : 36$
8. $54 : 48 : 60$
9. $18 : 9 : 27$
10. $25 : 15 : 20$
11. $24 : 18 : 15$
12. $100 : 75 : 125$

> Simplify the ratio $8\,\text{cm} : 12\,\text{cm} : 0.16\,\text{m}$
>
> $0.16\,\text{m} = 0.16 \times 100\,\text{cm} = 16\,\text{cm}$
>
> $8\,\text{cm} : 12\,\text{cm} : 0.16\,\text{m} = 8\,\text{cm} : 12\,\text{cm} : 16\,\text{cm}$
>
> $\phantom{8\,\text{cm} : 12\,\text{cm} : 0.16\,\text{m}} = 8 : 12 : 16$
>
> $\phantom{8\,\text{cm} : 12\,\text{cm} : 0.16\,\text{m}} = 2 : 3 : 4$ (dividing by 4)

Simplify the following ratios.

13. $40\,\text{cm} : 20\,\text{cm} : 10\,\text{cm}$
14. $24\,\text{p} : £1.20 : 72\,\text{p}$
15. $27\,\text{m} : 18\,\text{m} : 36\,\text{m}$
16. $15\,\text{km} : 25\,\text{km} : 10\,000\,\text{m}$
17. $90\,\text{p} : £1.10 : £1.30$
18. $48\,\text{mm} : 2\,\text{cm} : 40\,\text{mm}$
19. $25\,\text{p} : £1 : 75\,\text{p}$
20. $£2 : £1.20 : 90\,\text{p}$
21. $2\,\text{m} : 120\,\text{cm} : 150\,\text{cm}$
22. $700\,\text{m} : 1300\,\text{m} : \frac{1}{2}\,\text{km}$

In questions 23 and 24, several alternative answers are given. Write down the letter that corresponds to the correct answer.

23. $15 : 12 : 9$ is the same as

   A $3 : 4 : 5$ **B** $5 : 4 : 9$ **C** $5 : 4 : 3$ **D** $5 : 4 : 1$

24. $30 : 24 : 15$ is the same as

   A $10 : 6 : 5$ **B** $5 : 4 : 3$ **C** $10 : 8 : 5$ **D** $15 : 12 : 5$

DIVISION IN A GIVEN RATIO

If we are asked to divide a length of 5 cm into two parts so that the ratio of the lengths of the two parts is 2 : 3, it is easy to see that the two lengths must be 2 cm and 3 cm.

Now suppose that the length to be divided is 40 cm. We want the ratio of the two parts to be 2 : 3, so one part is made up of two portions and the other of 3 portions, i.e. there are 5 portions altogether.

Dividing the 40 cm length into 5 equal portions gives the length of one portion as $\frac{40}{5}$ cm = 8 cm,

so the length of the first part $= 2 \times 8$ cm

$= 16$ cm

and the length of the second part $= 3 \times 8$ cm

$= 24$ cm

(We can check by adding the two lengths, i.e. 24 cm + 16 cm = 40 cm)

EXERCISE 20i

Divide 60 kg into two parts in the ratio 7 : 5

There are (7 + 5) portions, i.e. 12 portions

1 portion $= \frac{60}{12}$ kg

$= 5$ kg

First part $= 7 \times 5$ kg

$= 35$ kg

Second part $= 5 \times 5$ kg

$= 25$ kg

(Check 25 kg + 35 kg = 60 kg)

1. Divide 60 cm into two parts in the ratio 5 : 1

2. Divide 90 p into two parts in the ratio 4 : 5

3. Divide 72 m into two parts in the ratio 2 : 7

4. Divide 18 kg into two parts in the ratio 1 : 2

5. Divide 14 m into two parts in the ratio 3 : 4

6. Divide 24 cm into two parts in the ratio 3 : 5

7. Divide 80 p between Eleanor and Mary in the ratio 5 : 3

8. Divide £1.20 between two people in the ratio 5 : 7

9. Divide 108 km into two parts in the ratio 1 : 8

10. Share 40 sweets between two people in the ratio 2 : 3

11. John is 18 years old and James is 15 years old. Share 66 p between them in the ratio of their ages.

12. Divide 80 m into three parts in the ratio 2 : 5 : 1

13. Divide 72 cm into three parts in the ratio 3 : 1 : 2

14. Divide 95 kg into three parts in the ratio 2 : 2 : 1

15. Divide 60 cm into three parts in the ratio 3 : 5 : 4

In questions 16 to 18, several alternative answers are given.
Write down the letter that corresponds to the correct answer.

16. When 20 cm is divided into two parts in the ratio 4 : 1, the two parts are

    A 5 cm, 1 cm   **B** 15 cm, 5 cm

    C 4 cm, 1 cm   **D** 16 cm, 4 cm

17. When 24 g is divided into two parts in the ratio 3 : 5, the two parts are

    A 3 g, 5 g   **B** 24 g, 40 g   **C** 15 g, 9 g   **D** 9 g, 15 g

18. When 36 g is divided into three parts in the ratio 4 : 2 : 3, the three parts are

    A 4 g, 2 g, 3 g   **B** 16 g, 8 g, 12 g

    C 16 g, 6 g, 14 g   **D** 36 g, 18 g, 27 g

PROBLEMS

EXERCISE 20j

> £30 is divided in the ratio $2 : 5 : 8$. Find the value of the smallest share.
>
> There are 15 portions
>
> 1 portion = £30 ÷ 15
>
> = £2
>
> Therefore the smallest share = £2 × 2
>
> = £4

1. A garden is divided into lawn, path and flower bed so that their areas are in the ratio $6 : 1 : 3$. If the area of the whole garden is $480 \, m^2$, find
   a) the area of the lawn
   b) the area of the path.

2. A metal alloy is composed of iron, copper and zinc in the ratio $10 : 4 : 1$. How much copper is there in 60 kg of the alloy?

3. There are 72 pupils in one year group and they are divided into three mathematics sets in the ratio $7 : 6 : 5$. How many pupils are there in the largest mathematics set?

4. Anne is 10 years old, Jane is 2 years older than Anne and Mary is 2 years older than Jane. A sum of £36 is divided amongst them in the ratio of their ages.
   a) What is the ratio of their ages?
   b) How much is Anne's share?

5. A T.V. station broadcasts from 07.30 to 23.30 hours. Times spent on news, on comedy programmes and on other broadcasts are in the ratio $2 : 5 : 9$.
   a) How many hours of news broadcasting were there?
   b) How many hours were spent on comedy programmes?

> A length of wood is cut into three parts in the ratio $2 : 3 : 7$. The longest part is 2.8 m in length. How long was the original piece of wood?
>
> There are 12 portions altogether.
>
> The longest part = 7 portions = 2.8 m
>
> Therefore 1 portion = $\dfrac{2.8}{7}$ m
>
> = 0.4 m
>
> The length of the original piece = 12 × 0.4 m
>
> = 4.8 m

6. A sum of money was shared between Tom and Harry in the ratio $3 : 4$. Tom received 48 p.
   a) What was the original sum of money?
   b) How much did Harry receive?

7. Three people shared out the contents of a box of chocolates in the ratio $4 : 5 : 3$. The smallest share consisted of 9 chocolates.
   a) How many chocolates were there in the box originally?
   b) How many chocolates were there in the largest share?

8. In a school choir, the numbers of singers from the first, second and third years are in the ratio $2 : 7 : 9$. There are 35 singers from the second year.
   a) How many singers are there from the three years altogether?
   b) How many singers are there from the first year?

9. The amounts of brown, beige and cream wool used in knitting a sweater are in the ratio $5 : 3 : 1$. The amount of beige wool used is 75 g.
   a) How much wool is used altogether?
   b) How much brown wool is used?

MAP SCALES

On a map let us suppose that 1 cm represents 1 km. Then the ratio of the distance on the map to the distance on the ground, i.e. the *real* distance, is 1 cm : 1 km.

$$\text{Now, } 1\,\text{km} = 1000\,\text{m}$$
$$= 100\,000\,\text{cm}$$
$$\text{so } 1\,\text{cm} : 1\,\text{km} = 1\,\text{cm} : 100\,000\,\text{cm}$$
$$= 1 : 100\,000$$

For this map, we say that the *map scale* or map ratio is $1 : 100\,000$.

On the cover of some ordnance survey maps you will find the ratio $1 : 50\,000$. This means that 1 cm represents 50 000 cm

i.e. $\qquad\qquad\qquad$ 1 cm represents $\frac{1}{2}$ km.

EXERCISE 20k

On a road atlas the map ratio is $1 : 200\,000$. What real distance is represented by 1 cm on the map? Give your answer in km.

The map ratio is $1 : 200\,000$

so 1 cm represents 200 000 cm

$$= \frac{200\,000}{100}\,\text{m} \qquad (100\,\text{cm} = 1\,\text{m})$$
$$= 2000\,\text{m}$$
$$= \frac{2000}{1000}\,\text{km} \qquad (1000\,\text{m} = 1\,\text{km})$$
$$= 2\,\text{km}$$

1. A map ratio is $1 : 5000$. What real distance in m is represented by 1 cm on the map?

2. A map ratio is $1 : 10\,000$. What real distance in m is represented by 1 cm on the map?

3. A map ratio is $1 : 10\,000\,000$. What real distance in km is represented by 1 cm on the map?

4. A map ratio is $1 : 20\,000$. What distance in m is represented by 1 cm on the map?

Ratio and Proportion

> A map ratio is 1 : 10 000. What real distance in metres is represented by 5 cm on the map?
>
> 1 cm represents 10 000 cm
>
> 5 cm represents 50 000 cm
>
> $= 500$ m
>
> $= \frac{1}{2}$ km

5. The ratio marked on a map is 1 : 5000. What real distance in m is represented by 10 cm on the map?

6. A map ratio is 1 : 100 000. What real distance in km is represented by 6 cm on the map?

7. The ratio on an ordnance survey map is 1 : 50 000. What distance in km is represented by 12 cm on the map?

8. On a street map of a town, the ratio is given as 1 : 1000. What distance is represented by 8 cm on the map?

9. A map ratio is 1 : 1 000 000. What distance in km is represented by 8 cm on the map?

> A map ratio is 1 : 50 000. What length on the map represents an actual distance of 4 km?
>
> 4 km = 400 000 cm
>
> 50 000 cm is represented by 1 cm
>
> Therefore 400 000 cm is represented by $\frac{400\,000}{50\,000}$ cm
>
> $= 8$ cm

10. The ratio on an ordnance survey map is 1 : 50 000. What length on the map represents an actual distance of 2.5 km?

11. A map ratio is 1 : 100 000. How many cms on the map represent a real distance of 7 km?

12. A map ratio is 1 : 10 000. What length on the map represents an actual distance of 800 m?

13. An ordnance survey map has a map ratio of 1 : 25 000. What length on the map represents a distance of 10 km?

DIRECT PROPORTION

"Proportion" is another word used for comparison. Two quantities are in proportion if they are always in the same ratio, e.g. if one quantity is doubled, so is the other one.

When we buy records at a fixed price each, the total cost of the records is proportional to the number bought, e.g. if we treble the number bought, we treble the total cost.

As another example, if a plane is flying at a steady speed then the distance travelled is proportional to the time taken.

EXERCISE 20I

> The total cost of 5 identical pens is 80 p.
>
> Find a) the cost of 1 pen   b) the cost of 7 pens.
>
> a) 5 pens cost 80 p   b) 7 pens cost 7 × 16 p
>
> ∴   1 pen costs $\frac{80 \text{ p}}{5}$ = 112 p
>
>   = 16 p   = £1.12

1. Six envelopes cost 54 p. Find
   a) the cost of 1 envelope
   b) the cost of 10 envelopes.

2. Four packets of cornflakes, all of equal mass, have a total mass of 2000 g.
   a) What is the mass of 1 packet?
   b) What is the mass of 9 packets?

3. If 8 m of chain costs 240 p, what is the cost of
   a) 1 m?
   b) 5 m?

4. Six labradors, all with equal appetites, eat 24 kg of food in a week.
   a) How much does one labrador eat in a week?
   b) How much do five of these labradors eat in a week?

Ratio and Proportion 271

5. Six men, working at the same rate, dig a trench 24 m long in a day.
   a) How long a trench would 1 man dig?
   b) How long a trench would 4 men dig?

6. Mr Jones finds he can buy 2 m of dowel for 24 p.
   a) What is the cost of 1 m?
   b) What is the cost of 5 m?

7. At a steady rate, a man drives 280 km in 4 hours.
   a) How far does he drive in 1 hour?
   b) How far does he drive in 3 hours?

In questions 8 to 16, you may need to find out about 1 item, as you did in questions 1 to 7, before completing your answer.

8. Nine pencils cost 72 p. What is the cost of 8 pencils?

9. A clock gains 15 minutes in 5 days. How much does it gain in 7 days?

10. At a steady rate, a man drives 72 miles in 2 hours. How far does he drive in 3 hours?

11. Six chairs cost £42. What is the cost of 5 of these chairs?

12. A clock gains 2 minutes in 6 days. How many days does it take to gain 5 minutes?

13. A carpet of area 12 m² costs £96. At the same cost per m², what is the cost of a carpet of area 15 m²?

14. Seven tickets for a concert cost £21. How much would 5 of these tickets cost?

15. If 3 m of shelving are needed for 135 copies of a text book, how many copies can be put on 5 m of shelving?

16. On 6 pages of printed text there are 2100 words. How many words are there on 8 pages?

MIXED EXERCISES

EXERCISE 20m
1. Simplify the ratio $54 : 36$
2. Simplify the ratio $\frac{3}{8} : 2$
3. Fill the gap in the ratio $10 : 8 = 1 :$
4. Divide 60 cm into 2 parts in the ratio $2 : 3$
5. Simplify the ratio $14 : 21 : 28$.
6. Mr and Mrs Lark have 6 grandsons and 4 granddaughters. Find the ratio of the number of grandsons to the number of granddaughters.
7. A map ratio is $1 : 10000$. Find the real distance in metres represented by 1 cm on the map.
8. 3 men, working at the same rate, build 15 m of wall. In the same time, how many metres of wall would 2 men build?

EXERCISE 20n
1. Simplify the ratio $30 : 36$
2. Simplify the ratio $2 : \frac{3}{4}$
3. Fill the gap in the ratio $6 : 5 = \ \ : 1$
4. Divide 72 p into two parts in the ratio $7 : 2$
5. Simplify the ratio $18 : 72 : 9$
6. John is 7 years old and Alan is 9 years old. Divide 80 p between them in the ratio of their ages.
7. An ordnance survey map is marked with the ratio $1 : 500\,000$. What distance in kilometres is represented by 7 cm on the map?
8. At a steady speed, a motorist drives 120 km in 3 hours. At the same speed, how far does he go in 5 hours?

EXERCISE 20p In the following questions, several alternative answers are given. Write down the letter that corresponds to the correct answer.

1. The ratio $24 : 16$ is the same as

    A $3 : 2$ **B** $2 : 3$ **C** $2 : 1$ **D** $1 : 3$

2. The ratio $\frac{2}{3} : 1$ is the same as

    A $3 : 2$ **B** $2 : 1$ **C** $2 : 3$ **D** $1 : 3$

3. The ratio $4:3$ is the same as

A $1:3$ **B** $1:0.75$ **C** $\frac{3}{4}:1$ **D** $4:1$

4. When 36 p is divided into 2 parts in the ratio $7:2$, the larger part is

A 7 p **B** 28 p **C** 8 p **D** 14 p

5. In a class of 30 pupils, 16 are boys. The ratio of boys to girls is

A $8:7$ **B** $8:15$ **C** $15:8$ **D** $7:8$

6. A map ratio is $1:5000$. The distance represented by 10 cm on the map is

A 10 km **B** 1 km **C** 5000 m **D** 500 m

21 INDICES AND APPROXIMATIONS

INDICES

Consider $2^3 \times 2^2$

$$2^3 \times 2^2 = 2 \times 2 \times 2 \times 2 \times 2 = 2^5$$

But if we add the powers on the LHS we get 5,

i.e. $\qquad 2^3 \times 2^2 = 2^{3+2} = 2^5$

This illustrates a general rule, i.e.

> a number to a power multiplied by the *same* number to another power can be simplified by adding the powers.

Now consider $2^5 \div 2^2$

$$2^5 \div 2^2 = \frac{\cancel{2} \times \cancel{2} \times 2 \times 2 \times 2}{\cancel{2} \times \cancel{2}} = 2^3$$

This time if we subtract the powers on the LHS we get 3,

i.e.
> a number to a power divided by the *same* number to another power can be simplified by subtracting the powers.

EXERCISE 21a Write as a single number in index form:

1. $2^3 \times 2^4$
2. $3^2 \times 3^3$
3. $5^2 \times 5^4$
4. $2^7 \div 2^3$
5. $4^5 \div 4^3$
6. $5^8 \div 5^4$
7. $3^7 \times 3^2$
8. $2^5 \div 2^4$
9. $7^2 \times 7^4$

ZERO AND NEGATIVE INDICES

Consider $2^3 \div 2^3$

$$2^3 \div 2^3 = \frac{2 \times 2 \times 2}{2 \times 2 \times 2} = 1$$

If we use the rule for dividing, we get $2^3 \div 2^3 = 2^{3-3} = 2^0$

This shows that 2^0 means 1.

Indices and Approximations 275

We will always get a zero index if we use the rule to divide a number by itself, i.e. $a^n \div a^n = a^0$.

When a number is divided by itself, the answer is 1.

Therefore
$$a^0 = 1$$
i.e. (any number)$^0 = 1$

Now consider $2^3 \div 2^4$

$$2^3 \div 2^4 = \frac{\cancel{2} \times \cancel{2} \times \cancel{2}}{\cancel{2} \times \cancel{2} \times \cancel{2} \times 2} = \frac{1}{2}$$

If we use the rule we get
$$2^3 \div 2^4 = 2^{3-4} = 2^{-1}$$

Therefore 2^{-1} means $\frac{1}{2}$

In the same way $4^2 \div 4^5 = \frac{\cancel{4} \times \cancel{4}}{\cancel{4} \times \cancel{4} \times 4 \times 4 \times 4} = \frac{1}{4^3}$

$$4^2 \div 4^5 = 4^{2-5} = 4^{-3}$$

Therefore 4^{-3} means $\frac{1}{4^3}$

These examples show that a negative sign in the index means "the reciprocal of",

i.e.
$$a^{-5} \text{ means "the reciprocal of } a^5\text{"}$$
so $$a^{-5} = \frac{1}{a^5}$$

EXERCISE 21b

Find the value of 6^{-1}

$$6^{-1} = \frac{1}{6^1} = \frac{1}{6}$$

Find the value of:

1. 2^{-1}
2. 4^{-1}
3. 10^{-1}
4. 3^{-1}
5. 9^{-1}
6. 12^{-1}
7. 7^{-1}
8. 100^{-1}

> Find the value of 5^{-2}
>
> $$5^{-2} = \frac{1}{5^2}$$
> $$= \frac{1}{25}$$

Find the value of:

9. 2^{-2}	12. 2^{-3}	15. 3^{-2}	**18.** 7^{-3}
10. 4^{-2}	13. 6^{-2}	16. 10^{-4}	**19.** 2^{-4}
11. 10^{-3}	14. 10^{-2}	17. 3^{-3}	**20.** 10^{-5}

> Find the value of $6^{-1} \times 2^2$
>
> $$6^{-1} \times 2^2 = \frac{1}{\cancel{6}_3} \times \cancel{4}^2$$
> $$= \frac{2}{3}$$

21. 9^2	24. 10^0	27. 100^{-1}	**30.** 8^0
22. $2^2 \times 3^2$	25. $2^3 \times 3^2$	28. $2^{-1} \times 3^{-1}$	**31.** 10^{10}
23. 10^6	26. 10^{-6}	29. 3^{-4}	**32.** $5^2 \times 2^2$

In questions 33 to 35 there are several alternative answers. Write down the letter that corresponds to the correct answer.

33. The value of 6^{-2} is

    A -36 **B** $\frac{1}{36}$ **C** -12 **D** $-\frac{1}{12}$

34. The value of $2^2 \times 3^{-1}$ is

    A -12 **B** $\frac{1}{12}$ **C** $\frac{1}{36}$ **D** $1\frac{1}{3}$

35. The value of 4^0 is

    A 4   **B** 1   **C** 0   **D** None of these

Indices and Approximations 277

STANDARD FORM (SCIENTIFIC NOTATION)

Very large numbers, or very small numbers, are more conveniently written in standard form. They take up less space and it is easier to compare sizes.

A number in standard form is written as a number between 1 and 10 multiplied by the appropriate power of 10.

For example 1.5×10^2 and 3.62×10^{-3} are in standard form, but 36.2×10^{-4} is not.

EXERCISE 21c

The following numbers are given in standard form. Write them as ordinary numbers.

a) 3.52×10^3 b) 1.04×10^{-3}

a) $\quad 3.52 \times 10^3 = 3.52 \times 1000$
$\qquad\qquad\quad\ = 3520$

b) $\quad 1.04 \times 10^{-3} = 1.04 \times \dfrac{1}{10^3}$
$\qquad\qquad\qquad\ = \dfrac{1.04}{1000}$
$\qquad\qquad\qquad\ = 0.00104$

Write as ordinary numbers:

1. 2.75×10^2
2. 1.3×10^{-2}
3. 4.26×10^3
4. 2.07×10^{-3}
5. 9.7×10^4
6. 7.5×10^{-1}
7. 9.8×10^5
8. 5.82×10^{-3}
9. 3.75×10^1
10. 2.92×10^{-4}
11. 8.04×10^2
12. 7.73×10^{-5}
13. 2.56×10^{-2}
14. 3.99×10^6
15. 4.00×10^{-3}

EXPRESSING NUMBERS IN STANDARD FORM

To write a number in standard form, first write the given figures as a number between 1 and 10. Then decide what power of 10 to multiply by to bring the number back to its correct size.

For example, to write 2520 in standard form we start with 2.52 which is a number between 1 and 10. To make 2.52 have the same size as 2520 we have to multiply it by 1000, or 10^3.

i.e. $\qquad\qquad 2520 = 2.52 \times 1000 = 2.52 \times 10^3$

EXERCISE 21d

Write 0.034 in standard form

$$0.034 = 3.4 \times \frac{1}{100}$$
$$= 3.4 \times 10^{-2}$$

Write the following numbers in standard form.

1. 273
2. 362
3. 42.3
4. 5600
5. 56
6. 0.79
7. 0.16
8. 0.025
9. 0.039
10. 0.0033
11. 502
12. 78.6
13. 0.94
14. 0.086
15. 757

16. 58000
17. 0.00029
18. 7600
19. 0.0038
20. 27250
21. 36940
22. 0.03994
23. 0.0004025
24. 372010
25. 0.50024
26. 788500
27. 2246
28. 87.32
29. 996.1
30. 55910

SCIENTIFIC CALCULATION AND STANDARD FORM

If you have a scientific calculator it will display very large or very small numbers in scientific notation. (Scientific notation is another name for standard form.) However only the power of 10 is given; 10 itself does not appear in the display.

Enter 0.00005 in your calculator

then press $\boxed{x^2}$ which gives $(0.00005)^2$

The display will read $\boxed{2.5 - 09}$

This means 2.5×10^{-9}

i.e. $(0.00005)^2 = 0.0000000025$

In some calculators the power is set higher, e.g. 2.5^{-09}

Next enter 730000 and press $\boxed{x^2}$

The display will read $\boxed{5.329\ 11}$ which means 5.329×10^{11}

i.e. $(730000)^2 = 532900000000$

Indices and Approximations 279

EXERCISE 21e Use your calculator to find the value of

1. $(36\,000)^2$
2. $(27\,500)^2$
3. $(8\,000\,000)^2$
4. $(239\,000)^2$
5. $(7800)^2$
6. $(5\,700\,000)^2$

7. $(0.00008)^2$
8. $(0.000064)^2$
9. $(0.000007)^2$
10. $(0.00009)^2$
11. $(0.00032)^2$
12. $(0.0000005)^2$

SIGNIFICANT FIGURES

If you were asked to measure the width of a table and give the answer correct to 2 d.p., your answer would depend on the unit that you used: it might be 0.75 metres or it could be 75.25 cm.

You could not give the answer in millimetres correct to two decimal places because it is impossible to measure the width of a table to the nearest $\frac{1}{100}$ th of a millimetre!

Therefore asking for a figure to be given correct to a number of decimal places is not always practical.

If you were asked for the width of the table correct to the first three digits, then your answer might be

$$0.753 \text{ m}$$

or
$$75.3 \text{ cm}$$

or
$$753 \text{ mm}$$

In all three cases we have the same degree of accuracy – i.e. the measurement is correct to the nearest mm, and we say that we have given the answer correct to *three significant figures*.

To find the 1st, 2nd, 3rd, ... significant figure in a number we can first write it in standard form. Then the digit to the left of the decimal point is the first significant figure, the next digit to the right is the second significant figure, ... and so on.

For example $\quad 0.007625 = 7.625 \times 10^{-3}$

where in 0.007625 the 1st s.f. is 7, 2nd s.f. is 6, 3rd s.f. is 2, 4th s.f. is 5; and in 7.625 the 1st s.f. is 7, 2nd s.f. is 6, 3rd s.f. is 2, 4th s.f. is 5.

To correct to a given number of significant figures we look at the next significant figure: if this is 5 or more we add 1 to the previous figure, if it is less than 5 we do not alter the previous figure.

For example, to correct 0.0070¦25 to two significant figures we look at the third significant figure.

In this case it is 2, so we do not alter the second significant figure,

i.e. $\qquad$ 0.007 025 = 0.0070 correct to 2 s.f.

EXERCISE 21f

Write down the third significant figure in 73.82 and state what it represents.

73.82 $\qquad$ (7.38¦2 × 10)

The 3rd s.f. is 8 and it represents tenths.

For each of the following numbers write down the significant figure indicated in brackets and state what it represents.

1.	8.273	(2nd)	**6.** 58.294	(4th)	**11.** 0.372	(2nd)
2.	827.3	(2nd)	**7.** 853.07	(4th)	**12.** 37.2	(3rd)
3.	0.08273	(3rd)	**8.** 20157	(3rd)	**13.** 37200	(4th)
4.	15.04	(2nd)	**9.** 0.005027	(3rd)	**14.** 0.0507	(2nd)
5.	1504	(3rd)	**10.** 370.74	(4th)	**15.** 507	(3rd)

Give the following numbers correct to 3 s.f.

a) 89472 $\qquad$ b) 0.089472

a) 894¦72 = 89500 correct to 3 s.f. $\qquad$ (8.94¦72 × 10^4)

b) 0.0894¦72 = 0.0895 correct to 3 s.f. $\qquad$ (8.94¦72 × 10^{-2})

Give the following numbers correct to 2 s.f.

16. 8.254   **19.** 12.8   **22.** 0.824
17. 4.073   **20.** 172   **23.** 0.0873
18. 2.16   **21.** 43.7   **24.** 0.00307

Give the following numbers correct to 3 s.f.

25. 9.267   **28.** 86.59   **31.** 0.02544
26. 6.058   **29.** 759.42   **32.** 0.87766
27. 4.397   **30.** 3699   **33.** 0.00050992

Give the following numbers correct to 4 s.f.

34. 8.5028   **37.** 58.924   **40.** 0.502937
35. 2.0193   **38.** 851.82   **41.** 0.0722384
36. 6.9237   **39.** 50456   **42.** 0.00990394

Give the following numbers correct to the number of significant figures indicated in the brackets.

43. 73.14 (2)   **46.** 70945 (4)   **49.** 1.2799 (3)
44. 0.05737 (3)   **47.** 0.009372 (2)   **50.** 0.788010 (4)
45. 889.4 (3)   **48.** 3.14159 (4)   **51.** 0.002657 (3)

USING A CALCULATOR

Calculators are marvellous aids because they take the drudgery out of many calculations. We cannot always be certain, however, that the answer they give is correct because *we* make mistakes when using them, particularly when we enter numbers. Our mistakes often result in outrageously incorrect answers so we should always ask "Is the answer reasonable?" One way to answer this question is to get a rough idea of the size of the answer that we expect. We can do this by correcting each number in the calculation to one significant figure and then working out the rough answer.

For example $257.38 \times 9.837 \approx 300 \times 10 = 3000$

On a calculator $257.38 \times 9.837 = 2531.8471$

This is the same sort of size as the rough estimate, so it is probably correct.

In this example we have written down all the figures shown in the display but this is not usually necessary. Answers are usually required correct to either three or four significant figures.

EXERCISE 21g

> Calculate 82.59×0.7326 giving the answer correct to 3 s.f.
>
> $(82.59 \times 0.7326 \approx 80 \times 0.7 = 56)$
>
> $82.59 \times 0.7326 = 60.50$ (writing down the first 4 s.f.)
>
> $\qquad\qquad\qquad = 60.5$ correct to 3 s.f.

First give a rough estimate for the following calculations, then use your calculator to give the answers correct to 3 s.f.

1. 27.8×5.243 **6.** 8.99×4.06
2. $57.2 \div 2.9$ **7.** $278 \div 27.3$
3. 8.742×0.2015 **8.** $26.9 \div 5.37$
4. $604 \div 58.2$ **9.** 67.3×0.92
5. 13.6×25.2 **10.** $55.5 \div 1.66$

11. 0.0278×6.34 **16.** $0.09374 \div 2.83$
12. 102.8×0.00792 **17.** 0.00572×8.65
13. $0.0392 \div 0.04273$ **18.** 372×953
14. 587.3×9046 **19.** $889 \div 9.07$
15. $(0.07332)^2$ **20.** 5055×2202

EXERCISE 21h In this exercise you are given several alternative answers. Write down the letter that corresponds to the correct answer.

1. The second significant figure of 825.3 represents

    A tens   **B** hundreds   **C** units   **D** tenths

2. Correct to two significant figures, the value of 0.796 is

    A 1.0   **B** 0.79   **C** 0.8   **D** 0.80

3. The value of $2^2 \times 2^3$ is

    A 64   **B** $\frac{1}{2}$ **C** 32   **D** 8

4. The value of $(0.04)^2$ is

    A 0.16   **B** 1.6   **C** 0.08   **D** 0.0016

5. A rough answer for 0.29×5.7 is

   A 2   **B** 5   **C** 0.3   **D** 6

6. The third significant figure of 80.53 represents

   A units   **B** tens   **C** hundredths   **D** tenths

7. $4^6 \div 4^3$ can be written as

   A 4^9 **B** 4^2 **C** 4^3 **D** 2

8. The value of 4^{-1} is

   A -4 **B** $-\frac{1}{4}$ **C** $\frac{1}{4}$ **D** 3

9. 0.872 in standard form is

   A 8.72×10^{-1} **B** 8.72×10

   C 8.72×10^2 **D** 8.72×10^{-2}

10. The value of 5.6×10^{-3} is

    A 0.56   **B** 0.056   **C** 0.0056   **D** 5600

22 THE TANGENT OF AN ANGLE

NAMING THE SIDES OF A RIGHT-ANGLED TRIANGLE

In Book 2 we gave names to each of the three sides in a right-angled triangle.

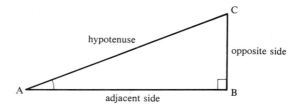

In triangle ABC, AC is the hypotenuse. It is the side opposite to the right angle.

For the angle A, BC is the opposite side and AB is the adjacent, or neighbouring, side.

EXERCISE 22a Copy the diagrams in questions 1 to 6.

In each diagram name the hypotenuse and the side opposite and adjacent to the marked angle.

1.
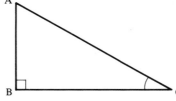

3.
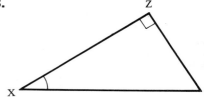

2.
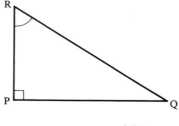

4.
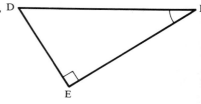

5. 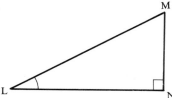 **6.**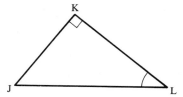

Questions 7 to 12 refer to the following diagrams.

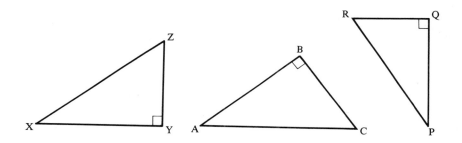

Copy and complete the following sentences.

7. In △XYZ, for angle X, XY is the .... side.

8. In △ABC, for angle A, AB is the .... side.

9. In △PQR, for angle R, PQ is the .... side.

10. In △XYZ, for angle Z, XY is the .... side.

11. In △PQR, for angle P, the opposite side is ....

12. In △ABC, the hypotenuse is ....

Questions 13 to 16 refer to the following diagram.

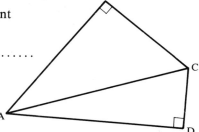

13. In △ABC, for B$\hat{A}$C, the adjacent side is ....

14. In △ACD, the hypotenuse is ....

15. In △ABC, for B$\hat{C}$A, AB is the ....

16. In △ACD, for C$\hat{A}$D, CD is the ....

TANGENT OF AN ANGLE

Draw the diagram given below as accurately as you can, following the instructions.

You may find it convenient to use squared paper.

Draw AX of any length and let $\hat{A}$ have a value between 20° and 40°.

Mark B, C and D so that AB = 5 cm, AC = 8 cm and AD = 10 cm.

Use a protractor or set square if necessary to draw BE, CF and DG perpendicular to AX as shown.

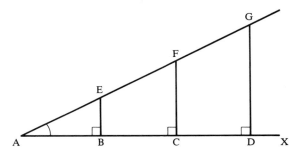

Measure BE, CF and DG.

Use these measurements to find the fractions $\dfrac{BE}{AB}$, $\dfrac{CF}{AC}$, and $\dfrac{DG}{AD}$ as decimals correct to 2 decimal places.

If we consider the angle A these fractions are given by $\dfrac{\text{the opposite side}}{\text{the adjacent side}}$ in triangles ABE, ACF and ADG.

The three fractions have the same value. This value is called the tangent of the angle A or briefly, tan $\hat{A}$.

Its size is stored, together with the tangents of other angles, in *natural tangent tables* and in some *calculators*.

Measure angle A. Using your calculator enter the size of $\hat{A}$, then press the button labelled "tan".

How does the value in the display compare with the values of the three fractions you have found above?

$$\tan \hat{A} = \frac{\text{opposite side}}{\text{adjacent side}}$$

EXERCISE 22b Use a calculator if you need one.

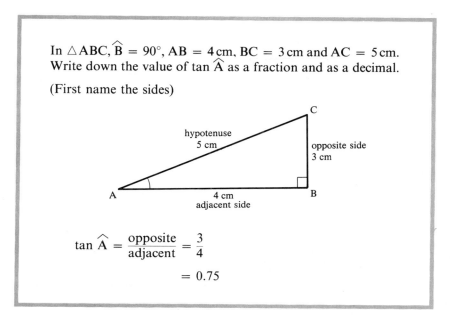

In △ABC, $\hat{B} = 90°$, AB = 4 cm, BC = 3 cm and AC = 5 cm. Write down the value of tan $\hat{A}$ as a fraction and as a decimal.

(First name the sides)

$$\tan \hat{A} = \frac{\text{opposite}}{\text{adjacent}} = \frac{3}{4}$$
$$= 0.75$$

In each of the following questions write down the tangent of the marked angle as a fraction and as a decimal (correct to 4 decimal places if necessary).

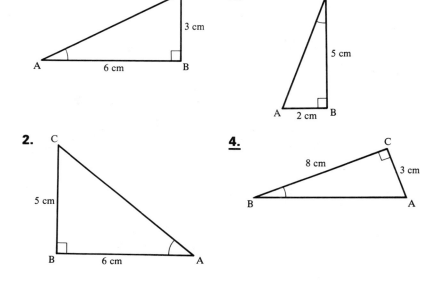

5.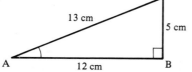

6.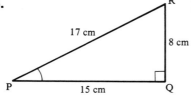

7.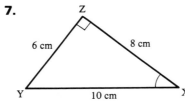

8.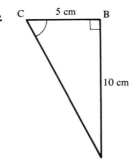

9.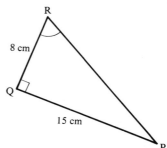

10.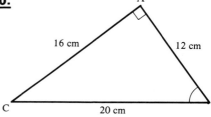

11.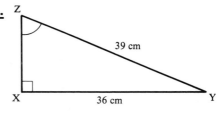

12.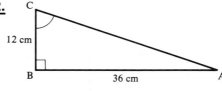

The Tangent of an Angle

For the following questions several alternative answers are given. Write down the letter that corresponds to the correct answer.

13.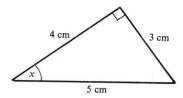

tan *x* is

A $\frac{3}{5}$ **B** $\frac{3}{4}$ **C** $\frac{4}{3}$ **D** $\frac{5}{3}$

14.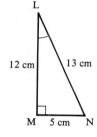

tan $\widehat{L}$ is

A $\frac{5}{12}$ **B** $\frac{5}{13}$ **C** $\frac{12}{5}$ **D** $\frac{13}{5}$

15.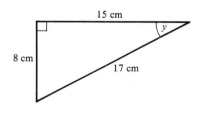

tan *y* is

A $\frac{17}{8}$ **B** $\frac{15}{17}$ **C** $\frac{8}{17}$ **D** $\frac{8}{15}$

16.

$\frac{AB}{AC}$ is

A tan $\widehat{A}$ **B** tan $\widehat{C}$ **C** tan $\widehat{B}$ **D** none of these

USING A CALCULATOR

To find the tangent of an angle, enter the size of the angle then press the "tan" button. Write the answer correct to 4 decimal places.

e.g. $\tan 36° = 0.7265$

Sometimes angles are given to the nearest tenth of a degree

e.g. $\tan 42.4° = 0.9131$

EXERCISE 22c Use a calculator to find the tangents of the following angles. Give values to 4 decimal places where necessary.

1. 56°	**7.** 78°	**13.** 32.4°
2. 44°	**8.** 45°	**14.** 3°
3. 59.6°	**9.** 60°	**15.** 52.1°
4. 82.1°	**10.** 30°	**16.** 66°
5. 37°	**11.** 17.6°	**17.** 15.4°
6. 26.8°	**12.** 68.7°	**18.** 49.2°

To find an angle when its tangent is given enter the value of the tangent and then press the inverse button followed by the tangent button. Write down the size of the angle correct to 1 decimal place.

EXERCISE 22d

Find the angle whose tangent is

a) 0.75   b) 0.447

a) $\tan \widehat{A} = 0.75$
    $\widehat{A} = 36.9°$

b) $\tan \widehat{B} = 0.447$
    $\widehat{B} = 24.1°$

Find the angles whose tangents are:

1. 0.5	**6.** 0.8	**11.** 0.2
2. 0.26	**7.** 0.44	**12.** 0.88
3. 0.873	**8.** 1	**13.** 0.345
4. 1.44	**9.** 1.64	**14.** 1.626
5. 1.736	**10.** 2.735	**15.** 3.668

FINDING AN ANGLE

EXERCISE 22e

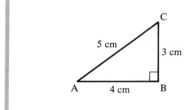

In triangle ABC, $\widehat{B} = 90°$, AB = 4 cm, BC = 3 cm and AC = 5 cm. Find $\widehat{A}$

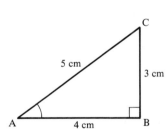

$$\tan \widehat{A} = \frac{\text{opposite side}}{\text{adjacent side}} = \frac{3}{4}$$
$$= 0.75$$
$$\widehat{A} = 36.9°$$

Use the information given in the diagrams to find $\widehat{A}$. Give your answers correct to 1 decimal place.

1. **2.**

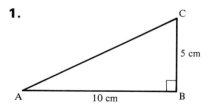

3.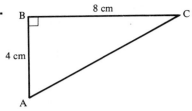

7.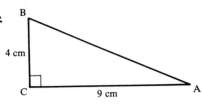

4.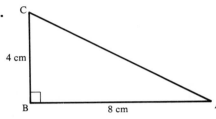

8.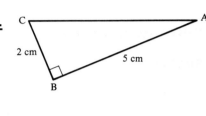

5.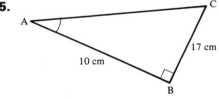

9.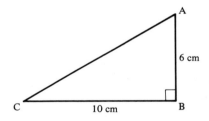

6.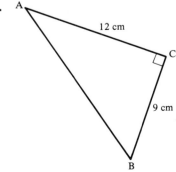

10.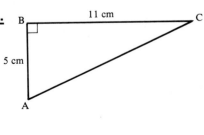

11.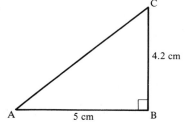

15.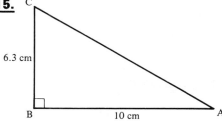

12.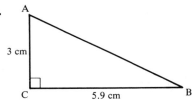

16.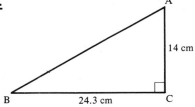

13.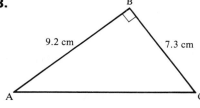

17.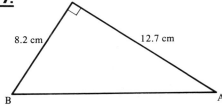

14.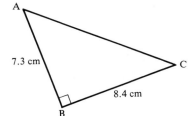

18.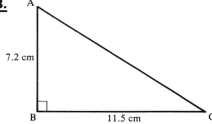

19. In triangle PQR, $\hat{P} = 90°$, QP = 8 cm and PR = 10 cm. Find $\hat{R}$.

20. In triangle DEF, $\hat{D} = 90°$, DE = 18 cm and DF = 21 cm. Find $\hat{E}$.

CALCULATIONS

When finding the tangent of an angle from a calculator write down its value as far as the fourth decimal place.

Write down all steps correct to four significant figures and give your final answer correct to three significant figures.

EXERCISE 22f

Use your calculator to find the value of
a) 12 tan 42°   b) 23 tan 62.3°

a) 12 tan 42° = 12 × 0.9004
  = 10.80
  = 10.8 (correct to 3 s.f.)

b) 26 tan 62.3° = 26 × 1.9047
  = 49.52
  = 49.5 (correct to 3 s.f.)

Use your calculator to find, correct to three significant figures, the value of:

1. 9 tan 37°
2. 13 tan 46°
3. 44 tan 62.3°
4. 27 tan 22.8°
5. 18 tan 53.4°

6. 8 tan 80°
7. 7 tan 34°
8. 34 tan 58.5°
9. 74 tan 32.3°
10. 58 tan 68.4°

11. 31.7 tan 23°
12. 13.5 tan 76°
13. 1.36 tan 34.6°
14. 72.4 tan 47.3°
15. 318 tan 82.1°

16. 14.8 tan 76°
17. 62.9 tan 57°
18. 3.49 tan 52.8°
19. 59.2 tan 32.6°
20. 247 tan 73.9°

FINDING THE OPPOSITE SIDE TO A GIVEN ANGLE

EXERCISE 22g

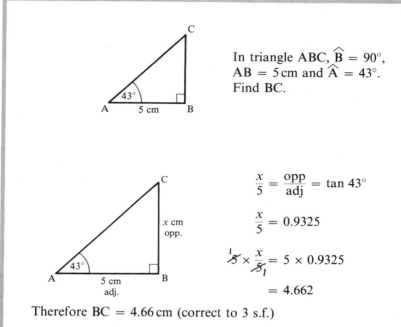

In triangle ABC, $\hat{B} = 90°$, AB = 5 cm and $\hat{A} = 43°$. Find BC.

$$\frac{x}{5} = \frac{\text{opp}}{\text{adj}} = \tan 43°$$

$$\frac{x}{5} = 0.9325$$

$$5 \times \frac{x}{5} = 5 \times 0.9325$$

$$= 4.662$$

Therefore BC = 4.66 cm (correct to 3 s.f.)

Use the information given in the diagram to find the required side correct to 3 s.f. First label the required side and the given side either adjacent or opposite.

1.
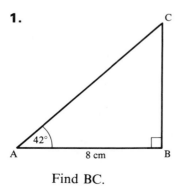

Find BC.

2.
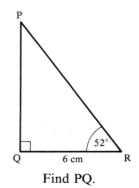

Find PQ.

3. Find AB.

4. Find XY.

5. Find AB.

6. Find EF.

7. Find PQ.

8. Find QR.

9. Find LM.

10. Find XY.

11. In triangle ABC, $\hat{B} = 90°$, $\hat{A} = 42°$ and AB = 8.2 cm. Find BC.

12. In triangle DEF, $\hat{D} = 90°$, $\hat{E} = 56°$ and DE = 31.4 cm. Find DF.

13. In triangle PQR, $\hat{R} = 90°$, $\hat{Q} = 12.4°$ and RQ = 28 cm. Find PR.

14. In triangle XYZ, $\hat{Z} = 90°$, $\hat{Y} = 65.4°$ and YZ = 6.3 cm. Find XZ.

15. In triangle ABC, $\hat{B} = 90°$, $\hat{C} = 37.3°$ and BC = 12.7 cm. Find AB.

FINDING THE ADJACENT SIDE FOR A GIVEN ANGLE

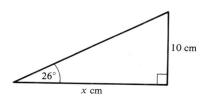

Sometimes the side whose length we are asked to find is adjacent to the given angle instead of opposite to it.

Using $\tan 26° = \dfrac{10}{x}$ gives an awkward equation so we work out the size of the other angle and use that angle instead.

In this case the other angle is 64°. (Remember that the three angles of a triangle add up to 180°.)

We now label the sides as "opposite" and "adjacent" to the angle of 64°.

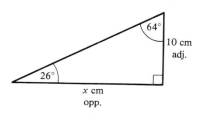

Using 64°, $\qquad \dfrac{x}{10} = \dfrac{\text{opp}}{\text{adj}} = \tan 64°$

so $\qquad \dfrac{x}{10} = 2.0503 \quad \text{giving} \quad x = 20.5 \quad \text{to 3 s.f.}$

EXERCISE 22h

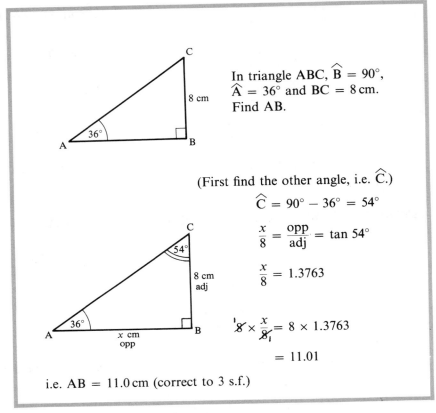

In triangle ABC, $\widehat{B} = 90°$, $\widehat{A} = 36°$ and BC = 8 cm. Find AB.

(First find the other angle, i.e. $\widehat{C}$.)

$\widehat{C} = 90° - 36° = 54°$

$\dfrac{x}{8} = \dfrac{\text{opp}}{\text{adj}} = \tan 54°$

$\dfrac{x}{8} = 1.3763$

$\overset{1}{\cancel{8}} \times \dfrac{x}{\cancel{8}_1} = 8 \times 1.3763$

$= 11.01$

i.e. AB = 11.0 cm (correct to 3 s.f.)

Use the information given in the diagram to find the required side correct to 3 s.f. It may first be necessary to find the third angle of the triangle.

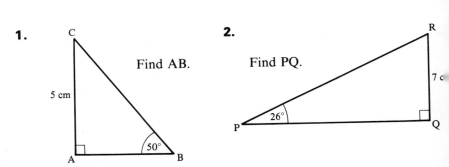

1. Find AB.

2. Find PQ.

3. Find XY.

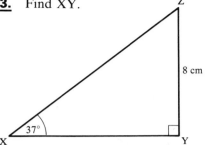

7. Find XZ.

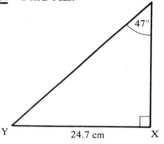

4. Find DF.

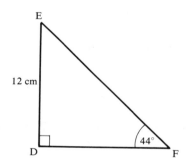

8. Find XZ.

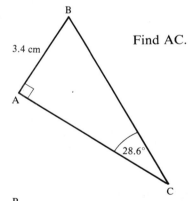

5. Find BC.

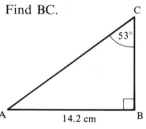

9. Find AC.

6. Find PQ.

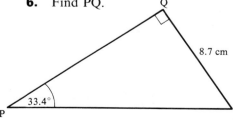

10. Find PQ.

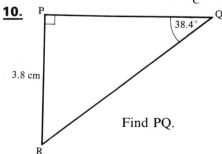

MIXED EXERCISES

EXERCISE 22i Questions 1 to 3 refer to the following diagram.

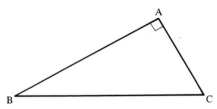

1. Name the side that is adjacent to $\widehat{B}$.

2. If AB = 10 cm and AC = 8 cm write down the value of $\tan \widehat{C}$.

3. If $\widehat{B} = 42°$, AC = 20 cm and AB = x cm, complete the following equation
$$\tan \ldots = \frac{x}{20}$$

4. Use your calculator to give the value of 12 tan 57° correct to three significant figures.

5.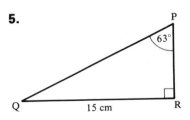

   Find PR giving your answer correct to three significant figures.

EXERCISE 22j

1. Write down the value of $\tan \widehat{C}$
   a) as a fraction
   b) as a decimal correct to 3 decimal places.

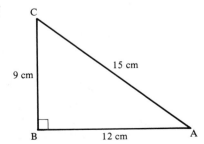

2. Use a calculator to find tan 37.7° correct to 4 decimal places.

3.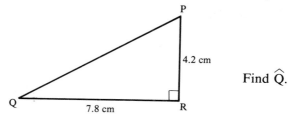

Find $\hat{Q}$.

4. Use your calculator to find, correct to three significant figures
   a) $12 \tan 72°$ b) $56.2 \tan 37.8°$

5. Find the angle whose tangent is   a) 0.7276   b) 1   c) 1.937.

EXERCISE 22k In this exercise several alternative answers are given. Write down the letter that corresponds to the correct answer.

1.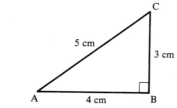

The tangent of the angle A is

A $\frac{4}{5}$ **B** $\frac{3}{5}$ **C** $\frac{4}{3}$ **D** $\frac{3}{4}$

2.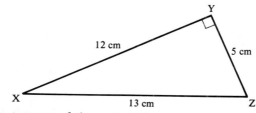

The tangent of the angle Z is

A $\frac{5}{13}$ **B** $\frac{5}{12}$ **C** $\frac{12}{5}$ **D** $\frac{12}{13}$

3.

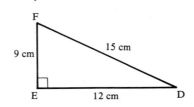

The tangent of angle D as a decimal fraction is

A 1.3333   **B** 0.75   **C** 0.8   **D** 0.6

4.

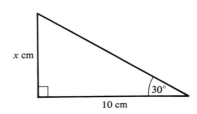

The value of x can be found from the equation

A $\dfrac{10}{x} = \tan 30°$ **B** $\dfrac{x}{10} = \tan 30°$

C $\dfrac{10}{x} = \tan 90°$ **D** $\dfrac{x}{10} = \tan 60°$

5.

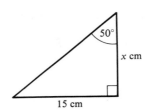

The value of x can be found from the equation

A $\dfrac{x}{15} = \tan 50°$ **B** $\dfrac{x}{15} = \tan 90°$

C $\dfrac{15}{x} = \tan 40°$ **D** $\dfrac{x}{15} = \tan 40°$

23 THE SINE AND COSINE OF AN ANGLE

In Chapter 22 we considered the fraction given by $\dfrac{\text{the opposite side}}{\text{the adjacent side}}$ in any right angled triangle.

We called this fraction the tangent of the angle.

i.e. $\tan \widehat{A} = \dfrac{\text{opp}}{\text{adj}}$

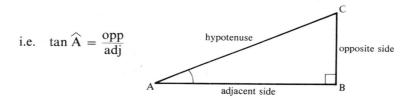

We shall now consider the fractions we get when the hypotenuse is used.

THE SINE OF AN ANGLE

Draw the diagram given below as accurately as you can, following the instructions. You may find it convenient to use squared paper.

Draw AX of any length and let $\widehat{A}$ have a value between 20° and 40°. Mark B, C and D so that AB = 5 cm, AC = 10 cm and AD = 15 cm.

Use a protractor or set square to draw the perpendiculars from B, C and D to AX as shown.

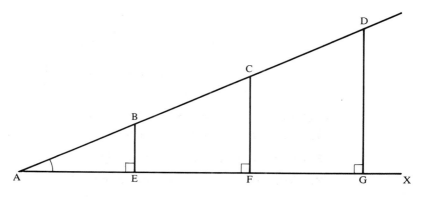

Measure BE, CF and DG.

303

Use these measurements to find the fractions $\frac{BE}{AB}$, $\frac{CF}{AC}$ and $\frac{DG}{AD}$ as decimals correct to two decimal places.

If we consider the angle A these fractions are given by $\frac{\text{the opposite side}}{\text{the hypotenuse}}$ in triangles ABE, ACF and ADG.

The three fractions have the same value. This value is called the sine of the angle A or briefly sin $\hat{A}$. Its size is stored, together with the sines of other angles, in natural sine tables and in some calculators.

Measure angle A in your diagram, and use a calculator to find sin $\hat{A}$. How does this value compare with the values of the three fractions found above?

$$\sin \hat{A} = \frac{\text{opposite side}}{\text{hypotenuse}}$$

All values found from the calculator should be written down correct to 4 significant figures and answers should be given correct to 3 significant figures unless stated otherwise.

The sizes of angles should be given correct to 1 decimal place.

EXERCISE 23a Questions 1 to 4 refer to the following diagram.

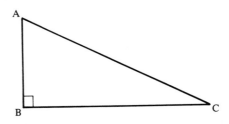

Copy the following sentences and fill in the blanks.
1. For angle C, the opposite side is ..... and the hypotenuse is .....
2. Side AB is the ..... side for angle A.
3. Side BC is the ..... side for angle C.
4. For angle A, BC is the ..... and AC is the ....

In $\triangle ABC$, $\widehat{B} = 90°$, $AC = 20$ cm, $AB = 16$ cm and $BC = 12$ cm. Write down the value of $\sin \widehat{A}$ as a fraction and as a decimal.

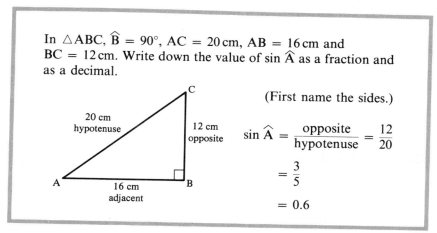

(First name the sides.)

$$\sin \widehat{A} = \frac{\text{opposite}}{\text{hypotenuse}} = \frac{12}{20}$$
$$= \frac{3}{5}$$
$$= 0.6$$

In each of the following questions write down the sine of the marked angle as a fraction and as a decimal (correct to 4 decimal places if necessary).

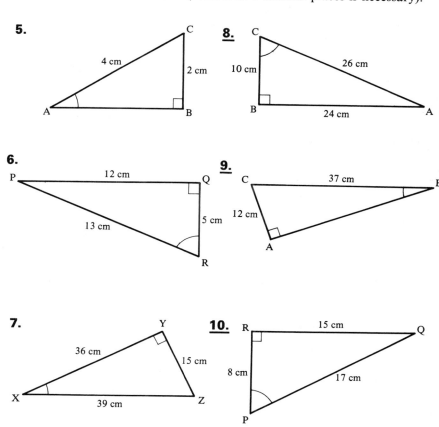

11. Use the diagram to complete the following ratios:

a) $\sin \widehat{A} = \dfrac{\quad}{AC}$

b) $\sin \widehat{C} = \dfrac{\quad}{AC}$

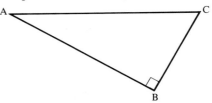

USING A CALCULATOR

To find the sine of an angle, enter the size of the angle then press the "sin" button. Write the answer correct to 4 decimal places.

e.g. $\quad\quad\quad\quad\quad\quad\quad\quad \sin 63° = 0.8910$

and $\quad\quad\quad\quad\quad\quad\quad \sin 39.3° = 0.6334$

EXERCISE 23b Use a calculator to find the sines of the following angles. Give values correct to 4 decimal places where necessary.

1. 44°
2. 67°
3. 54.7°
4. 18.2°
5. 80°
6. 33°
7. 17.4°
8. 59.8°
9. 56°
10. 49°
11. 82.5°
12. 72.3°

To find the angle when the sine is given enter the value of the sine and then press the inverse button followed by the sin button. Write down the size of the angle correct to 1 decimal place.

e.g. $\quad\quad\quad\quad\quad$ if $\sin \widehat{A} = 0.7264$

$\quad\quad\quad\quad\quad\quad\quad\quad\quad \widehat{A} = 46.6°$

EXERCISE 23c Find the angles whose sines are:

1. 0.4264
2. 0.1297
3. 0.8143
4. 0.5645
5. 0.7359
6. 0.2431
7. 0.8844
8. 0.5649
9. 0.2645
10. 0.3782
11. 0.9271
12. 0.6677

FINDING AN ANGLE

EXERCISE 23d

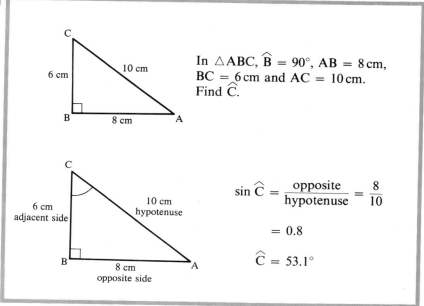

In △ABC, $\hat{B} = 90°$, AB = 8 cm, BC = 6 cm and AC = 10 cm. Find $\hat{C}$.

$$\sin \hat{C} = \frac{\text{opposite}}{\text{hypotenuse}} = \frac{8}{10}$$

$$= 0.8$$

$$\hat{C} = 53.1°$$

Use the information given in the diagram to find $\hat{A}$.

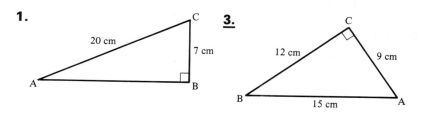

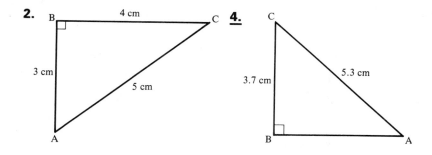

5.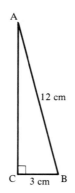

7.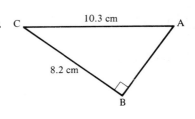

6.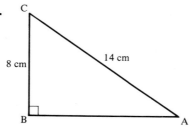

8.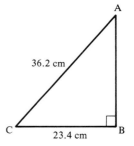

CALCULATIONS

EXERCISE 23e

Find 54.7 sin 48.8°

$$54.7 \sin 48.8° = 54.7 \times 0.7524$$
$$= 41.15$$
$$= 41.2 \text{ correct to 3 s.f.}$$

Use your calculator to find, correct to 3 significant figures, the value of:

1. 10 sin 34°
2. 36 sin 59°
3. 88 sin 48.5°
4. 4.2 sin 61.3°
5. 15 sin 62°
6. 82 sin 16°
7. 57 sin 70.9°
8. 5.9 sin 32.5°
9. 20 sin 85°
10. 59 sin 72°
11. 18 sin 62.6°
12. 15.6 sin 50.7°

The Sine and Cosine of an Angle

FINDING THE OPPOSITE SIDE FOR A GIVEN ANGLE

EXERCISE 23f

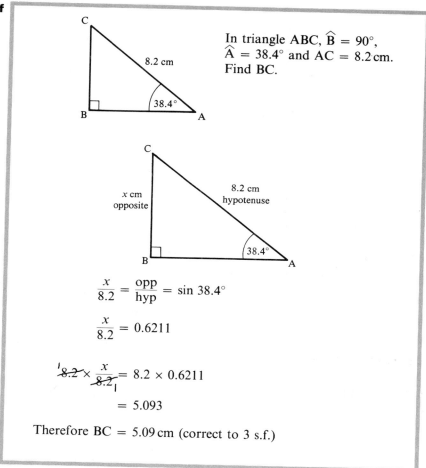

In triangle ABC, $\widehat{B} = 90°$, $\widehat{A} = 38.4°$ and AC = 8.2 cm. Find BC.

$$\frac{x}{8.2} = \frac{\text{opp}}{\text{hyp}} = \sin 38.4°$$

$$\frac{x}{8.2} = 0.6211$$

$$8.2 \times \frac{x}{8.2} = 8.2 \times 0.6211$$

$$= 5.093$$

Therefore BC = 5.09 cm (correct to 3 s.f.)

Use the information given in the diagram to find the required side.

1. Find BC.

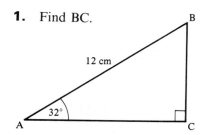

2. Find PQ.

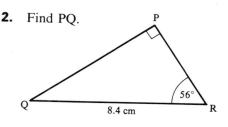

3. Find YZ.

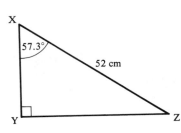

5. Find PR.

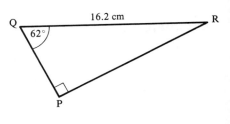

4. Find AB.

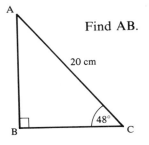

6. Find XY.

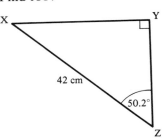

7. In △ABC, $\hat{B} = 90°$, $\hat{C} = 37.4°$ and AC = 16.4 cm. Find AB.

8. In △PQR, $\hat{Q} = 90°$, $\hat{R} = 61.8°$ and PR = 37.9 cm. Find PQ.

THE COSINE OF AN ANGLE

Use the diagram you drew from the instructions given on page 303.

Measure AE, AF and AG.

Find the fractions $\dfrac{AE}{AB}$, $\dfrac{AF}{AC}$ and $\dfrac{AG}{AD}$.

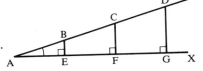

If we consider the angle A in triangles ABE, ACF and ADG, these ratios are the ratios of the adjacent side to the hypotenuse.

The three fractions have the same value. This value is called the cosine of the angle A or briefly cos $\hat{A}$. Its size is stored, together with the cosines of other angles, in natural cosine tables and in some calculators.

Measure angle A in your diagram, and use a calculator to find cos $\hat{A}$. How does this value compare with the three fractions you found above?

$$\cos \hat{A} = \frac{\text{adjacent side}}{\text{hypotenuse}}$$

EXERCISE 23g Questions 1 to 4 refer to the following diagram.

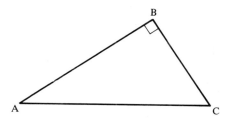

Copy the following sentences and fill in the blanks.

1. For angle A the adjacent side is ..... and the hypotenuse is .....

2. Side BC is the ..... side for angle C.

3. Side AC is the .....

4. For angle C, BC is the ..... and AC is the .....

In each of the following questions write down the cosine of the marked angle as a fraction and as a decimal (correct to 4 d.p. if necessary).

5.
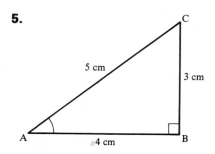

7.
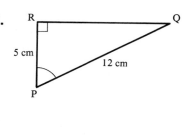

6.
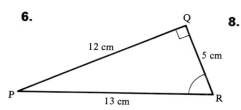

8.
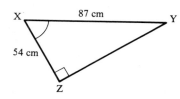

9.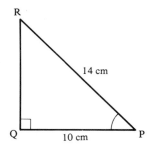

11.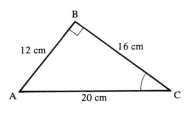

10.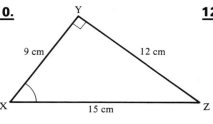

12.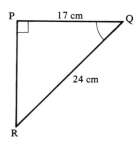

13. Use a calculator to find the cosines of the following angles.
   a) 68°
   b) 23.9°
   c) 42.7°
   d) 49°
   e) 59.2°
   f) 18.4°
   g) 27°
   h) 17.6°
   i) 73.6°

14. Use a calculator to find the angles whose cosines are
   a) 0.7464
   b) 0.2668
   c) 0.5374
   d) 0.8283
   e) 0.5492
   f) 0.9106
   g) 0.1479
   h) 0.1696
   i) 0.7009

In $\triangle ABC$, $\widehat{B} = 90°$, $BC = 7\,\text{cm}$ and $AC = 12\,\text{cm}$. Find $\widehat{C}$.

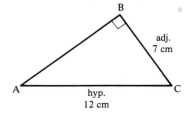

$$\cos \widehat{C} = \frac{\text{adj}}{\text{hyp}} = \frac{BC}{AC}$$

$$= \frac{7}{12}$$

$$= 0.5833$$

$$\widehat{C} = 54.3°$$

The Sine and Cosine of an Angle   313

Use the information given in the diagram to find $\hat{A}$.

15.
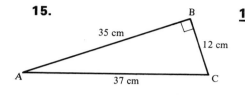

18.
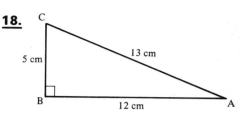

16.

19.

17.

20.
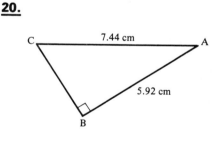

CALCULATIONS

EXERCISE 23h Use your calculator to find, correct to 3 significant figures, the value of:

1. 20 cos 42°
2. 6.71 cos 63.4°
3. 4.97 cos 14.7°
4. 73 cos 60°
5. 34.2 cos 82.8°
6. 15.3 cos 15.7°
7. 59 cos 49°
8. 80.2 cos 58.4°
9. 57.9 cos 32.5°

FINDING THE ADJACENT SIDE FOR A GIVEN ANGLE

EXERCISE 23i

In triangle ABC, $\widehat{B} = 90°$, AC = 24 cm and $\widehat{A} = 33°$
Find AB.

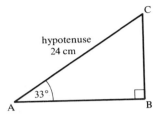

$$\frac{x}{24} = \frac{\text{adj}}{\text{hyp}} = \cos 33°$$

Therefore $\dfrac{x}{24} = 0.8387$

$$\cancel{24} \times \frac{x}{\cancel{24}} = 24 \times 0.8387$$

$$= 20.13$$

Therefore AB = 20.1 cm (correct to 3 s.f.)

Use the information given in the diagram to find the required side.

1. Find BC.

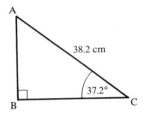

4. Find AC.

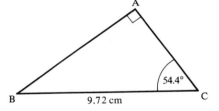

2. Find EF.

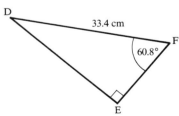

5. Find PQ.

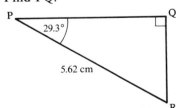

3. Find XY.

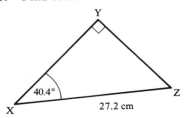

6. Find RT.

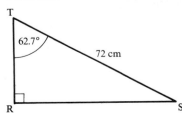

SUMMARY

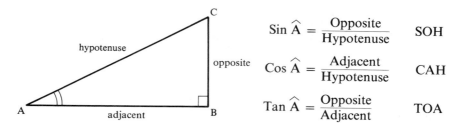

$$\text{Sin } \hat{A} = \frac{\text{Opposite}}{\text{Hypotenuse}} \quad \text{SOH}$$

$$\text{Cos } \hat{A} = \frac{\text{Adjacent}}{\text{Hypotenuse}} \quad \text{CAH}$$

$$\text{Tan } \hat{A} = \frac{\text{Opposite}}{\text{Adjacent}} \quad \text{TOA}$$

Some people find this easier to remember by thinking of SOHCAHTOA or a sentence like

"Some old hangars can almost hold two old aeroplanes"

MIXED QUESTIONS

EXERCISE 23j This exercise uses all three ratios: sine, cosine and tangent. To decide which ratio to use, label the sides.

1. Find AB.

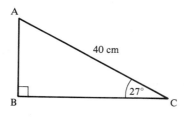

3. Find PQ.

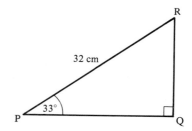

2. Find $\hat{A}$.

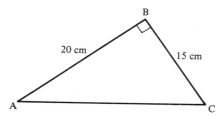

4. Find XZ.

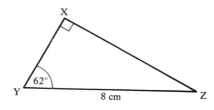

5. Find BC.

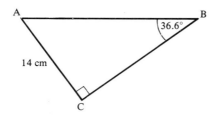

6. Find LM.

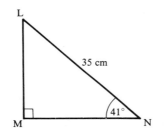

7. Find $\widehat{A}$.

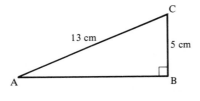

8. Find $\widehat{R}$.

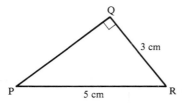

9. Find AB.

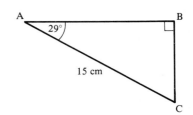

10. Find $\widehat{Q}$.

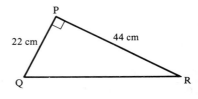

11. Find RS.

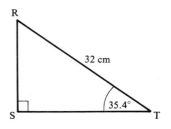

12. Find XZ.

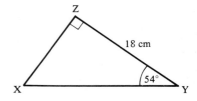

The Sine and Cosine of an Angle 317

13. Find RQ.

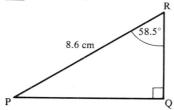

15. Find $\hat{N}$.

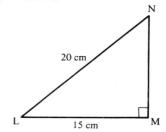

14. Find AB.

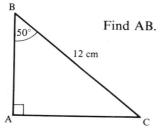

16. Find $\hat{Z}$.

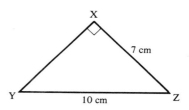

EXERCISE 23k In this exercise several alternative answers are given. Write down the letter that corresponds to the correct answer.

1.

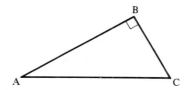

The cosine of $\hat{A}$ is

A $\dfrac{AC}{AB}$ **B** $\dfrac{BC}{AC}$ **C** $\dfrac{AB}{AC}$ **D** $\dfrac{AB}{BC}$

2.

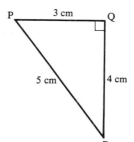

The sine of $\hat{R}$ is

A $\dfrac{4}{5}$ **B** $\dfrac{3}{5}$ **C** $\dfrac{3}{4}$ **D** $\dfrac{5}{3}$

3.

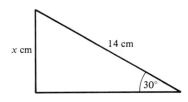

The value of x can be found from the equation

A $\dfrac{x}{14} = \tan 30°$ **B** $\dfrac{14}{x} = \cos 30°$

C $\dfrac{x}{14} = \sin 60°$ **D** $\dfrac{x}{14} = \sin 30°$

4.

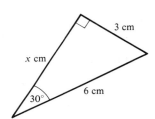

The value of x can be found from the equation

A $\dfrac{6}{x} = \cos 30°$ **B** $\dfrac{x}{3} = \tan 30°$

C $\dfrac{x}{6} = \cos 60°$ **D** $\dfrac{x}{6} = \cos 30°$

5.

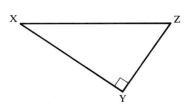

The tangent of $\widehat{Z}$ is

A $\dfrac{XY}{XZ}$ **B** $\dfrac{YZ}{XY}$ **C** $\dfrac{XY}{YZ}$ **D** $\dfrac{YZ}{XY}$

6.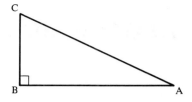

In the diagram $\frac{BC}{CA}$ is equal to

A $\tan \widehat{A}$ **B** $\sin \widehat{C}$ **C** $\cos \widehat{A}$ **D** none of these

24 ALGEBRAIC PRODUCTS

BRACKETS

Remember that $\quad 3(x+7) = 3x + 21$

and that $\quad 3(2x-5) = 6x - 15$

EXERCISE 24a Multiply out:

1. $4(x+2)$
2. $7(x-3)$
3. $5(3x+2)$
4. $6(5x-4)$
5. $5(x+3)$
6. $4(x-5)$
7. $7(2x+7)$
8. $3(3x-2)$
9. $2(x+7)$
10. $3(3x-4)$
11. $5(7x-2)$
12. $4(x-8)$

EXERCISE 24b

Multiply out a) $3x(x+4)$ b) $4a(b-2c)$

a) $\quad 3x(x+4) = 3x^2 + 12x$

b) $\quad 4a(b-2c) = 4ab - 8ac$

Multiply out:

1. $5x(x+2)$
2. $4x(x-5)$
3. $7x(2x-5)$
4. $4a(3b-2c)$
5. $3a(2b+3c)$
6. $6x(x+3)$
7. $3x(x-7)$
8. $2x(3x-4)$
9. $5a(b+3c)$
10. $7a(3b-2c)$
11. $4x(x+5)$
12. $10x(x+2)$

PRODUCT OF TWO BRACKETS

Sometimes we wish to remove both brackets in an expression such as $(x+3)(x+5)$ i.e. we need to multiply $(x+3)$ by $(x+5)$. To do this we multiply *every term* in the first bracket by *every term* in the second bracket.

$$(x+3)(x+5) = x \times x \ + \ x \times 5 \ + \ 3 \times x \ + \ 3 \times 5$$
$$= x^2 + 5x + 3x + 15$$
$$= x^2 + 8x + 15$$

i.e. $\quad (x+3)(x+5) = x^2 + 8x + 15$

320

Algebraic Products 321

We say that $x^2 + 8x + 15$ is the *expansion* of $(x + 3)(x + 5)$.
Conversely if we *expand* $(x + 3)(x + 5)$ we get $x^2 + 8x + 15$.

You will be more likely to get the correct answer each time if you always multiply the brackets together in the same order.

We suggest:
1) the first terms in the brackets
2) the outside terms
3) the inside terms
4) the second terms in the brackets

Thus $(x + 7)(x + 5) = x^2 + 5x + 7x + 35$

i.e. $(x + 7)(x + 5) = x^2 + 12x + 35$

EXERCISE 24c Expand:

1. $(x + 2)(x + 1)$
2. $(x + 10)(x + 5)$
3. $(x + 7)(x + 11)$
4. $(a + 4)(a + 9)$
5. $(m + 6)(m + 3)$
6. $(x + 4)(x + 3)$
7. $(x + 7)(x + 8)$
8. $(x + 3)(x + 9)$
9. $(b + 5)(b + 8)$
10. $(p + 8)(p + 4)$

Expand $(x - 5)(x - 7)$

$(x - 5)(x - 7) = x^2 - 7x - 5x + 35$
$= x^2 - 12x + 35$

Expand:

11. $(x - 1)(x - 4)$
12. $(x - 6)(x - 3)$
13. $(x - 2)(x - 10)$
14. $(a - 4)(a - 7)$
15. $(p - 8)(p - 3)$
16. $(x - 3)(x - 5)$
17. $(x - 7)(x - 5)$
18. $(x - 4)(x - 8)$
19. $(b - 1)(b - 9)$
20. $(t - 2)(t - 8)$

> Expand $(x + 2)(x - 5)$
>
> $$(x + 2)(x - 5) = x^2 - 5x + 2x - 10$$
> $$= x^2 - 3x - 10$$

Expand:

21. $(x + 3)(x - 4)$

22. $(x + 4)(x - 5)$

23. $(x - 2)(x + 6)$

24. $(a - 5)(a + 7)$

25. $(p + 8)(p - 2)$

26. $(x - 4)(x + 5)$

27. $(x + 4)(x - 6)$

28. $(x - 3)(x + 7)$

29. $(b - 8)(b + 1)$

30. $(q + 2)(q - 5)$

In questions 31 to 33 several alternative answers are given. Write down the letter that corresponds to the correct answer.

31. The expansion of $(x - 7)(x - 12)$ is

A $x^2 - 5x + 19$ **B** $x^2 - 19x + 84$

C $x^2 - 19x - 84$ **D** $x^2 + 5x + 84$

32. The expansion of $(x + 4)(x - 9)$ is

A $x^2 - 5x + 36$ **B** $x^2 + 5x - 36$

C $x^2 - 5x - 36$ **D** $x^2 + 5x + 36$

33. The expansion of $(x + 5)(x + 8)$ is

A $x^2 + 40x + 40$ **B** $x^2 + 13x + 13$

C $x^2 + 13x - 40$ **D** $x^2 + 13x + 40$

FINDING THE PATTERN

You may have noticed in the previous exercise that when you expanded the brackets and simplified the answers, there was a definite pattern linking the numbers in the given brackets and the numbers in the answer.

e.g. $(x + 5)(x + 7) = x^2 + 7x + 5x + 35$
$$= x^2 + 12x + 35$$

We could have written this

$$(x + 5)(x + 7) = x^2 + (+5 +7)x + (5) \times (7)$$
$$= x^2 + 12x + 35$$

Algebraic Products 323

Similarly $(x - 3)(x + 6) = x^2 + (-3 + 6)x + (-3) \times (6)$
$$= x^2 + 3x - 18$$
and $(x - 4)(x - 7) = x^2 + (-4 - 7)x + (-4) \times (-7)$
$$= x^2 - 11x + 28$$

In each case the coefficient of x^2 is 1.
Multiplying the two numbers in the brackets gives the number term in the expansion. *Collecting* the two numbers in the brackets gives the number of xs in the expansion.

EXERCISE 24d

In the expansion of $(x + 9)(x - 5)$
a) collect the two numbers   b) multiply the two numbers

$$(x + 9)(x - 5)$$
a) Collecting the two numbers gives
$$(+9) + (-5) = 9 - 5 = 4$$
b) Multiplying the two numbers gives
$$(+9) \times (-5) = -45$$

For the following expansions:
a) collect the two numbers   b) multiply the two numbers.

1. $(x + 4)(x + 6)$
2. $(x + 3)(x + 1)$
3. $(x - 9)(x - 3)$
4. $(x - 5)(x - 2)$
5. $(x + 7)(x - 5)$

6. $(x + 4)(x + 7)$
7. $(x + 6)(x + 1)$
8. $(x - 2)(x - 12)$
9. $(x + 9)(x - 2)$
10. $(x - 4)(x + 5)$

Expand $(x + 5)(x - 12)$

$$(x + 5)(x - 12) = x^2 + (5 + (-12))x + (5) \times (-12)$$
$$\text{collecting the}\text{multiplying}$$
$$\text{numbers}\text{the numbers}$$
$$= x^2 - 7x - 60$$

Use the idea of a pattern, as illustrated, to expand the following products.

11. $(x + 7)(x - 3)$

12. $(x - 11)(x + 5)$

13. $(x + 4)(x - 6)$

14. $(x - 8)(x + 3)$

15. $(x - 10)(x + 2)$

16. $(x + 8)(x - 10)$

17. $(x + 1)(x - 8)$

18. $(x - 10)(x + 12)$

In questions 19 to 22 several alternative answers are given. Write down the letter that corresponds to the correct answer.

19. When $(x + 7)$ is multiplied by $(x - 4)$ the x term is

    A $3x$ **B** $-3x$ **C** $-11x$ **D** $11x$

20. When $(x + 4)$ is multiplied by $(x - 5)$ the x term is

    A $-20x$ **B** $-x$ **C** $9x$ **D** $-9x$

21. When $(x - 7)$ is multiplied by $(x + 7)$ the x term is

    A $-14x$ **B** 0 **C** $49x$ **D** $14x$

22. When $(x + 10)$ is multiplied by $(x + 12)$ the x term is

    A $22x$ **B** $120x$ **C** $2x$ **D** $-22x$

The pattern is similar when the brackets are less simple.

EXERCISE 24e

Expand $(3x + 1)(x + 2)$

$$(3x + 1)(x + 2) = 3x^2 + 6x + x + 2$$
$$= 3x^2 + 7x + 2$$

Expand the following products.

1. $(2x + 1)(x + 1)$

2. $(5x + 2)(x + 4)$

3. $(3x + 1)(3x + 2)$

4. $(5x + 6)(2x + 3)$

5. $(x + 1)(3x + 1)$

6. $(4x + 3)(x + 5)$

7. $(4x + 1)(x + 7)$

8. $(3x + 5)(2x + 1)$

Algebraic Products

> Expand $(3x - 2)(x - 2)$
>
> $$(3x - 2)(x - 2) = 3x^2 - 6x - 2x + 4$$
> $$= 3x^2 - 8x + 4$$

Expand:

9. $(2x - 1)(x - 1)$
10. $(4x - 3)(x - 2)$
11. $(7x - 2)(x - 5)$
12. $(3x - 1)(3x - 4)$
13. $(5x - 1)(x - 3)$
14. $(3x - 2)(x - 4)$
15. $(4x - 5)(x - 2)$
16. $(4x - 3)(4x - 1)$

> Expand $(3x + 2)(2x - 3)$
>
> $$(3x + 2)(2x - 3) = 6x^2 - 9x + 4x - 6$$
> $$= 6x^2 - 5x - 6$$

Expand:

17. $(2x + 1)(3x - 2)$
18. $(5x - 4)(2x + 3)$
19. $(7x - 1)(2x + 3)$
20. $(3x + 5)(4x - 1)$
21. $(5x + 2)(3x - 4)$
22. $(3x - 2)(4x + 5)$
23. $(2x + 5)(5x - 2)$
24. $(7x - 2)(3x + 1)$
25. $(7x + 1)(9x + 2)$
26. $(2x - 9)(2x + 1)$
27. $(3x - 4)(2x - 7)$
28. $(8x + 5)(3x + 1)$
29. $(3x - 7)(4x + 3)$
30. $(7x - 8)(2x - 3)$

In questions 31 and 32 several alternative answers are given. Write down the letter that corresponds to the correct answer.

31. The expansion of $(4x - 1)(3x - 2)$ is

A $12x^2 + 14x - 2$
B $12x^2 - 7x + 2$
C $12x^2 - 14x + 2$
D $12x^2 - 11x + 2$

32. The expansion of $(3x - 4)(2x + 1)$ is

A $5x^2 - 5x + 4$
B $6x^2 + 5x - 4$
C $5x^2 - 11x - 4$
D $6x^2 - 5x - 4$

In questions 33 and 34 several alternative answers are given. Write down the letter that corresponds to the correct answer.

33. When $(5x - 3)$ is multiplied by $(3x + 5)$ the x term is

    A $-34x$ **B** $-16x$ **C** $16x$ **D** 0

34. When $(7x + 2)$ is multiplied by $(3x - 7)$ the x term is

    A $-55x$ **B** $-43x$ **C** $+43x$ **D** $+7x$

IMPORTANT PRODUCTS

1. $(x + a)^2 = (x + a)(x + a)$
   $= x^2 + xa + ax + a^2$
   $= x^2 + 2ax + a^2$ (xa is the same as ax)

   i.e. $(x + a)^2 = x + 2ax + a^2$

   so $(x + 4)^2 = x + 8x + 16$

2. $(x - a)^2 = (x - a)(x - a)$
   $= x^2 - xa - ax + a^2$
   $= x^2 - 2ax + a^2$

   i.e. $(x - a)^2 = x^2 - 2ax + a^2$

   so $(x - 3)^2 = x^2 - 6x + 9$

You should learn these results thoroughly, for they occur time and time again. Given the left hand side you should be able to write down the right hand side, and given the right hand side you should be able to express it in the form given on the left hand side.

EXERCISE 24f Expand:

1. $(x + 2)^2$ **4.** $(x + 5)^2$ **7.** $(x + 4)^2$

2. $(x + 1)^2$ **5.** $(x + 3)^2$ **8.** $(x + 8)^2$

3. $(x + 7)^2$ **6.** $(x + 9)^2$ **9.** $(x + 10)^2$

10. $(a + 5)^2$ **12.** $(b + 2)^2$ **14.** $(c + 8)^2$

11. $(b + 4)^2$ **13.** $(a + 9)^2$ **15.** $(y + 11)^2$

Expand:

16. $(x-3)^2$
17. $(x-7)^2$
18. $(x-9)^2$
19. $(x-5)^2$
20. $(x-2)^2$

21. $(x-10)^2$
22. $(x-6)^2$
23. $(x-11)^2$
24. $(x-1)^2$
25. $(x-8)^2$

Expand $(2x+5)^2$

$$(2x+5)^2 = (2x)^2 + 2(2x)(5) + (5)^2$$
i.e. $$(2x+5)^2 = 4x^2 + 20x + 25$$

Expand:

26. $(2x+3)^2$
27. $(3x+2)^2$
28. $(5x+3)^2$
29. $(6x+1)^2$
30. $(7x+5)^2$

31. $(2x+1)^2$
32. $(4x+3)^2$
33. $(5x+2)^2$
34. $(6x+5)^2$
35. $(3x+7)^2$

Expand $(3x-5)^2$

$$(3x-5)^2 = (3x)^2 + 2(3x)(-5) + (-5)^2$$
i.e. $$(3x-5)^2 = 9x^2 - 30x + 25$$

Expand:

36. $(2x-5)^2$
37. $(5x-2)^2$
38. $(9x-4)^2$
39. $(3x-1)^2$
40. $(2x-3)^2$

41. $(6x-5)^2$
42. $(4x-3)^2$
43. $(3x-4)^2$
44. $(7x-2)^2$
45. $(8x-5)^2$

DIFFERENCE BETWEEN TWO SQUARES

EXERCISE 24g

> Expand a) $(x + 2)(x - 2)$ b) $(2x + 5)(2x - 5)$
>
> a) $(x + 2)(x - 2) = x^2 - 2x + 2x - 4$
> $= x^2 - 4$
>
> b) $(2x + 5)(2x - 5) = 4x^2 - 10x + 10x - 25$
> $= 4x^2 - 25$

Expand:

1. $(x + 5)(x - 5)$

2. $(x + 7)(x - 7)$

3. $(x - 6)(x + 6)$

4. $(x + 12)(x - 12)$

5. $(x + 4)(x - 4)$

6. $(x + 3)(x - 3)$

7. $(x - 10)(x + 10)$

8. $(x + 8)(x - 8)$

9. $(2x + 1)(2x - 1)$

10. $(3x + 2)(3x - 2)$

11. $(8x + 5)(8x - 5)$

12. $(4x - 1)(4x + 1)$

13. $(5x + 1)(5x - 1)$

14. $(4x + 7)(4x - 7)$

15. $(7x + 3)(7x - 3)$

16. $(6x - 5)(6x + 5)$

17. $(5 + 3m)(5 - 3m)$

18. $(5y - 4)(5y + 4)$

19. $(8y + 5)(8y - 5)$

20. $(5p - 2)(5p + 2)$

In questions 21 to 23 several alternative answers are given. Write down the letter that corresponds to the correct answer.

21. The expansion of $(x + 4)(x - 4)$ is

A $x^2 - 8x + 16$ **B** $x^2 - 16$

C $x^2 + 16$ **D** $x^2 - 8x - 16$

22. The expansion of $(2x + 5)(2x - 5)$ is

A $4x^2 - 20x + 25$ **B** $4x^2 - 20x - 25$

C $4x^2 - 25$ **D** $4x^2 + 25$

23. The expansion of $(7x - 3)(7x + 3)$ is

A $49x^2 - 42x - 9$ **B** $49x^2 - 42x + 9$

C $49x^2 - 21x - 9$ **D** $49x^2 - 9$

Algebraic Products 329

The results from this exercise are very important, especially when written the other way around.

i.e. $\qquad a^2 - b^2 = (a + b)(a - b) \quad \text{or} \quad (a - b)(a + b)$

We refer to this as "factorising the difference between two squares".

SUMMARY

The following is a summary of the most important examples found in this chapter that will be required in future work.

1. $5(2x + 1) = 10x + 5$
2. $(x + 4)(x + 5) = x^2 + 9x + 20$
3. $(x - 3)(x - 7) = x^2 - 10x + 21$
4. $(x + 5)(x - 6) = x^2 - x - 30$
5. $(3x + 1)(4x + 3) = 12x^2 + 13x + 3$
6. $(2x - 3)(3x - 1) = 6x^2 - 11x + 3$
7. $(3x + 2)(2x - 3) = 6x^2 - 5x - 6$

Note that
a) if the signs in the brackets are the same, i.e. both + or both −, then the number term is positive (see examples 2, 3, 5 and 6 above)

whereas
b) if the signs in the brackets are different, i.e. one + and one −, then the number term is negative (see examples 4 and 7 above).

c) The middle term is given by collecting the products of the outside terms in the brackets and the inside terms in the brackets.

Most important of all we must remember the general expansions.

$$(x + a)^2 = x^2 + 2ax + a^2$$
$$(x - a)^2 = x^2 - 2ax + a^2$$
$$(x + a)(x - a) = x^2 - a^2$$

MIXED EXERCISES

EXERCISE 24h Expand:

1. $7(x - 5)$
2. $4x(x + 8)$
3. $5x(2x - 7)$
4. $(x + 10)(x + 7)$
5. $(x - 4)(x - 9)$
6. $(x + 8)^2$
7. $(x + 8)(x - 8)$
8. $(x - 7)(x + 1)$
9. $(5x + 2)(x - 3)$
10. $(7x + 8)(2x + 3)$
11. $(7x - 8)(3x - 2)$
12. $(7x - 8)(3x + 2)$
13. $(5x - 7)^2$
14. $(4x + 9)(4x - 9)$

EXERCISE 24i Expand:

1. $(x - 5)^2$
2. $(2x - 9)^2$
3. $9(2x - 3)$
4. $(x + 7)(x + 9)$
5. $(5x + 1)(5x - 1)$
6. $8x(3x - 7)$
7. $(7x + 11)(3x - 2)$
8. $(2x + 7)(2x - 7)$
9. $(x + 11)^2$
10. $(5x + 3)^2$
11. $5x(x - 4)$
12. $(8x + 9)(3x + 2)$
13. $(x - 7)(x + 8)$
14. $(x + 4)(x - 1)$

EXERCISE 24j Expand:

1. $(a + 1)(a - 2)$
2. $6p(3p - 1)$
3. $(2p - 3)(2p + 3)$
4. $(b + 4)^2$
5. $(5p + 2)(7p - 4)$
6. $t(3t - 4)$
7. $(3b + 10)^2$
8. $5a(a + 2)$
9. $(m - 4)(m + 8)$
10. $8(m - 4)$
11. $(a - 7)^2$
12. $(5p - 2)^2$
13. $(4a - 3)(5a + 2)$
14. $(5r + 3)(3r + 5)$

EXERCISE 24k In this exercise several alternative answers are given. In each question write down the letter that corresponds to the correct answer.

1. The x^2 term in the expansion of $(2x - 1)(3x + 5)$ is

   A $5x^2$ **B** $7x^2$ **C** $6x^2$ **D** x^2

2. The number term in the expansion of $(x - 4)(x + 4)$ is

    A 16   **B** -16 **C** 8   **D** -8

3. The x term in the expansion of $(x - 6)(x + 4)$ is

    A $10x$ **B** $2x$ **C** $-24x$ **D** $-2x$

4. The x term in the expansion of $(2x - 1)(3x + 1)$ is

    A $-x$ **B** 0 **C** $5x$ **D** $-2x$

HARDER EXAMPLES

EXERCISE 24I

> Expand $(4 - x)(2 + 3x)$
>
> $$(4 - x)(2 + 3x) = 8 + 12x - 2x - 3x^2$$
> $$= 8 + 10x - 3x^2$$

Expand:

1. $(5 + x)(2 + x)$ **4.** $(6 - 5x)(4 - 3x)$

2. $(5 - 2x)(3 + x)$ **5.** $(4 - 3x)(5 + x)$

3. $(4 - 3x)(1 + 2x)$ **6.** $(5 - x)(4 - 3x)$

> Expand $(4 + x)(x - 2)$
>
> $$(4 + x)(x - 2) = 4x - 8 + x^2 - 2x$$
> $$= x^2 + 2x - 8$$

Expand:

7. $(3 + x)(x + 7)$ **10.** $(3 + 2x)(1 - 2x)$

8. $(5 - 3x)(3x + 1)$ **11.** $(3 - 2x)(1 - 2x)$

9. $(2 - x)(5x - 1)$ **12.** $(3 - 2x)(1 + 2x)$

> Expand $(3x + 2y)(x + y)$
>
> $(3x + 2y)(x + y) = 3x^2 + 3xy + 2yx + 2y^2$
> $= 3x^2 + 5xy + 2y^2$

13. $(2x + 3y)(x + y)$
14. $(5x + 2y)(2x + y)$
15. $(7x + 2y)(3x + 4y)$
16. $(4x + 3y)(x + y)$
17. $(6x + 5y)(x + 3y)$
18. $(5x + 3y)(4x + 5y)$

> Expand $(4x - 3y)^2$
>
> $(4x - 3y)^2 = 16x^2 - 24xy + 9y^2$

Expand:

19. $(5x + y)^2$
20. $(2a + 7b)^2$
21. $(4p + 3q)^2$
22. $(4x + 5y)^2$
23. $(7b - 5c)^2$
24. $(3r - 7s)^2$

> Expand $(2x + 3y)(2x - 3y)$
>
> $(2x + 3y)(2x - 3y) = 4x^2 - 6xy + 6xy + 9y^2$
> $= 4x^2 - 9y^2$

25. $(5x + 2y)(5x - 2y)$
26. $(3x + 5y)(3x - 5y)$
27. $(4x - 3y)(4x + 3y)$
28. $(2a + 3b)(2a - 3b)$
29. $(6s - 5t)(6s + 5t)$
30. $(5y - 4z)(5y + 4z)$

25 MEAN, MODE AND MEDIAN

REPRESENTATIVE NUMBERS

When we have a large set of numbers, it is sometimes useful to have one number that we can use as a "typical member" so that it can represent the set. There are three such representative numbers, the mean, the mode, and median. We will look at each in turn.

MEAN OR AVERAGE

The mean of a set of numbers is the average as found in Chapter 12. Therefore the mean is the sum of the numbers divided by however many of them there are. For example, for the numbers 2, 4, 5, 7, 10

the mean is
$$\frac{2 + 4 + 5 + 7 + 10}{5}$$
$$= 28 \div 5$$
$$= 5.6$$

The mean is the most commonly used representative number.

EXERCISE 25a Find the mean value of the following sets of numbers.

1. 3, 5, 7, 3, 7

2. 2, 4, 2, 7, 5

3. 10, 12, 16, 14, 12, 8

4. 21, 32, 25, 27, 36, 27

5. 4, 2, 1, 7, 3, 5, 6

6. 7, 10, 9, 12, 15, 17, 14

7. 1.3, 1.5, 1.2, 1.0, 1.5

8. 2.5, 1.9, 2.6, 1.2, 0.3

9. 0.15, 0.02, 0.91, 1.02, 1.67, 0.01

10. 2.7, 8.3, 15.6, 7.3, 12.2, 15.9, 30.4

MODE

The mode is the number that occurs most often in a given set.

For example, for the numbers 2, 2, 4, 4, 4, 5, 6, 6 the mode is 4 because 4 occurs three times and no other number occurs more than twice.

If the numbers in a set are all different, there is no mode.

For example, there is no mode for the set of numbers 1, 2, 3, 5, 10, 12.

There can be two (or more) modes in a set. Consider, for example, the set of numbers 1, 2, 2, 3, 4, 5, 5, 6, 7; both 2 and 5 occur twice so both 2 and 5 are modes.

The mode of a set is sometimes a more suitable representative number than the mean. For example, suppose that the manager of a shoe shop keeps a record of the number of pairs of shoes that are sold of each size, and finds that there is a mode of, say, size $6\frac{1}{2}$. This information tells him to order more pairs in this size than in any other size.

The mode is easier to find when the numbers are arranged in order of size.

EXERCISE 25b Write down the mode, or modes, of the following sets of numbers. If there is no mode, write "none".

1. 3, 4, 5, 6, 6, 7, 8, 9
2. 1, 4, 6, 6, 7, 8, 9, 9, 9, 10
3. $2\frac{1}{2}$, $2\frac{1}{2}$, 3, $3\frac{1}{2}$, 4, $4\frac{1}{2}$, $4\frac{1}{2}$, 5, $5\frac{1}{2}$, $5\frac{1}{2}$, $5\frac{1}{2}$, 6, 6, 7
4. 8, 10, 12, 15, 17, 25, 30
5. 3, 8, 8, 10, 15, 15, 19, 20
6. 2, 4, 6, 8, 10, 10, 12, 12, 14, 16
7. 0.1, 0.2, 0.3, 0.3, 0.4, 0.5, 0.5, 0.6, 0.6
8. 10, 12, 13, 13, 13, 14, 15, 16, 16, 19, 19, 19, 20, 20
9. 1.2, 2.4, 3.6, 4.8, 6.0, 7.2
10. 2, 2, 3, 3, 4, 4, 5, 5, 6, 6

Remember to arrange these numbers in order of size before finding the mode.

11. 3, 8, 5, 7, 3, 2, 6, 9
12. 10, 9, 8, 9, 12, 15, 7, 9, 8

13. 1.6, 0.7, 2.5, 1.6, 0.8

14. 10, 9, 10, 8, 7, 6, 9, 10

15. 0.8, 1.2, 0.6, 1.3, 0.8, 0.5

16. 1.9, 0.7, 2.6, 3.7, 6.2, 0.5

17. $1\frac{1}{2}$, 1, 2, $1\frac{1}{2}$, 3, $2\frac{1}{2}$, 3, 1, $1\frac{1}{2}$, 3

18. 0.2, 0.1, 0.3, 0.5, 0.7, 0.4, 0.6

19. 26, 35, 30, 27, 30, 31, 26, 35

20. 1, 2, 5, 2, 1, 4, 3, 2, 5, 1, 2

MEDIAN

When we arrange a set of numbers in order of size, the median is the number in the middle. For example, for the seven numbers

$$2, 3, 4, ⑥, 8, 10, 12$$

the median is 6.

When there is an even number of numbers in the set, the median is the average of the two middle numbers. For example, for the six numbers

$$1, 2, \underline{4, 6}, 9, 10$$

the median is the average of 4 and 6, i.e. $\frac{4+6}{2} = 5$

For some sets of numbers the median is a better representative number than the mean. For example, in the end of term exams, Peter's marks were 5, 52, 59, 60, 68.

The mean mark is 48.8. This is lower than all his marks except one and it does not give a fair impression of his performance. This is because the low mark of 5 has pulled down Peter's average. It would be much fairer in this case to give the median, which is 59, as a representative mark.

EXERCISE 25c Find the median of the following sets of numbers. Remember to arrange numbers in order of size when necessary.

1. 1, 2, 4, 5, 8

2. 2, 3, 4, 5, 6

3. 2, 5, 2, 3, 4, 6, 3

4. 8, 2, 10, 5, 7, 3

5. 10, 20, 50, 51, 53, 54, 56

6. 8.2, 9.4, 10.0, 11.6, 12.2

7. 3, 6, 9, 15, 17, 21, 27, 29

8. 2.4, 3.6, 5.2, 0.4, 7.3, 7.9

9. 3, 2, 4, 3, 9

10. 3, 7, 4, 2, 5, 4

11. 10, 18, 12, 14, 16, 19

12. 2, 7, 3, 4, 7, 9, 8, 10

13. 1, 9, 3, 2, 12, 10, 4, 6

14. 1.2, 0.7, 1.5, 1.1, 2.2, 1.4

15. 28, 52, 10, 19, 29, 56, 84, 3

FINDING THE MEDIAN OF A LARGER SET

It is easy to find the median of a small set of numbers.

For a large set, say of n numbers arranged in order of size, the median is the $\left(\dfrac{n+1}{2}\right)$th number.

For 29 numbers, for example, the median is the $\left(\dfrac{29+1}{2}\right)$th number, i.e. the 15th number.

For 30 numbers, the median is the $\left(\dfrac{30+1}{2}\right)$th number, i.e. the $15\frac{1}{2}$th number. This means the average of the 15th and 16th numbers.

To find the median of the set of numbers

$$3, 4, 5, 3, 7, 8, 3, 3, 6, 2, 4, 6, 4, 6, 3, 13, 4, 3, 3, 2$$

we first arrange the numbers in size order.

$$2, 2, 3, 3, 3, 3, 3, 3, 3, 4, 4, 4, 4, 5, 6, 6, 6, 7, 8, 13$$

There are 20 numbers in this set so the median is the $\left(\dfrac{20+1}{2}\right)$th number

i.e. the $10\frac{1}{2}$th number

which is the average of the 10th and 11th numbers.

Counting from the left shows that the 10th number is 4

and the 11th number is 4

therefore the median is the average of 4 and 4 which is 4.

Mean, Mode and Median 337

EXERCISE 25d

Find the median of the set of numbers

0.1, 2.5, 3.7, 2.4, 1.1, 1.2, 0.2, 1.6, 1.4, 2.7, 3.8, 7.2

(Arrange the numbers in order of size.)

0.1, 0.2, 1.1, 1.2, 1.4, 1.6, 2.4, 2.5, 2.7, 3.7, 3.8, 7.2

The median is the $\left(\frac{12+1}{2}\right)$ th value.

i.e. the $6\frac{1}{2}$ th value

$= $ average of 1.6 and 2.4

$= \frac{1.6 + 2.4}{2}$

$= \frac{4}{2}$

$= 2$

Find the median of each of the following sets of numbers.

1. 1, 1, 2, 2, 3, 3, 3, 4, 5, 6, 7, 7, 8, 9, 10, 10, 10

2. 2, 3, 3, 4, 5, 5, 6, 7, 9, 10, 10, 11, 11, 15, 16, 17, 19, 20, 25

3. 1.2, 1.3, 1.5, 1.7, 1.9, 2.4, 3.6, 3.9, 4.0, 4.2, 4.7

4. 1, 3, 9, 15, 17, 20, 24, 32, 42, 43, 45, 47, 49, 52

5. 2, 3, 1, 5, 7, 2, 4, 3, 9, 8, 4, 7, 3, 9, 2, 1, 5

6. 2.7, 3.6, 1.9, 3.2, 0.8, 1.7, 3.9, 4.2, 5.3

MIXED QUESTIONS

EXERCISE 25e Find the mean, median and mode (when there is one) of the following sets of numbers.

1. 1, 3, 5, 5, 6, 7, 8

2. 2, 2, 3, 4, 5, 6, 6

3. 1, 2, 3, 4, 5

4. 1, 1.1, 1.2, 1.3, 1.4, 1.5

5. 7, 4, 9, 3, 2, 5

6. 2.4, 0.9, 1.7, 3.8, 1.7

Each of questions 7, 8 and 9 is followed by several alternative answers. Write down the letter that corresponds to the correct answer.

7. The median of 7, 2, 4, 7, 5 is

    A 2   **B** 3   **C** 5   **D** 7

8. The mean of 1, 5, 4, 2, 7, 5 is

    A 4   **B** 5   **C** 6   **D** $4\frac{1}{2}$

9. The mode of 1, 7, 1, 2, 4, 2, 1 is

    A $2\frac{4}{7}$ **B** 2   **C** 7   **D** 1

10. Ten music students took a grade 3 piano examination. They obtained the following marks.

    106, 125, 132, 140, 108, 102, 75, 135, 146, 123

Find the mean and median marks.

Which of these two representative marks would be most useful to the teacher who entered the students?

(Give *brief* reasons; do not write an essay on the subject.)

11. The first eight customers at a supermarket one Saturday spent the following amounts.

    £25.10, £3.80, £20.50, £15.70, £38.40, £9.60, £46.20, £10.50

Find the mean and median amounts spent.

12. A small firm employs ten people. The salaries of the employees are as follows.

    £30,000, £8,000, £5,000, £5,000, £5,000, £5,000,
    £5,000, £4,000, £3,000, £1,500

Find the mean, mode and median salary. Which of these three figures is a trade union official likely to be interested in, and why?

13. Thirty 15 year olds were asked how much pocket money they received each week and the following amounts (in pence) were recorded.

    0, 0, 0, 50, 50, 50, 100, 100, 100, 100, 100, 100, 100, 150, 150, 200, 200,
    200, 200, 200, 200, 200, 200, 200, 200, 250, 250, 250, 500, 1000

Find the mean, mode and median amounts. Which of the three measures would you use when asking for more pocket money, and why?

Mean, Mode and Median

FINDING THE MODE FROM A FREQUENCY TABLE

A set of numbers such as those given in question 13 of the last exercise is not always written out. It is more usual to find them given in a frequency table.

For example, 10 fifteen year olds were asked how much pocket money they received each week. The following frequency table shows the number of children who receive each listed sum of money.

Amount of pocket money (pence)	0	50	100	150	200
Frequency (number of children)	1	2	3	2	2

It is very easy to find the mode from a frequency table. All we have to do is to identify the number which occurs most often, i.e. the number with the highest frequency. We can see immediately that more pupils get 100 pence than any other amount, i.e. 100 pence occurs most often. Therefore the mode is 100 pence.

EXERCISE 25f In each of the following questions find the mode from the frequency table.

1. The number of prickles on each of 50 holly leaves were counted. The table shows the results.

Number of prickles	1	2	3	4	5	6
Frequency	4	2	8	7	20	9

2. A six-sided dice was thrown 50 times. The table gives the number of times each score was obtained.

Score	1	2	3	4	5	6
Frequency	7	8	10	8	5	12

3. Three coins were tossed together 30 times and the number of heads per throw was recorded.

Number of heads	0	1	2	3
Frequency	3	12	10	5

4. Cars passing a school entrance between 9 and 9.30 one morning were observed and the number of people per car was recorded. The table shows this information.

Number of people	1	2	3	4	5	6
Frequency	50	38	15	6	4	7

5. An express checkout in a supermarket is for people with 8 items or less. A survey was carried out and the results are shown in the table.

Number of items	1	2	3	4	5	6	7	8
Frequency	10	5	20	35	15	42	50	80

6. This table shows, for one particular Saturday, the number of hours that 40 children spent watching television.

Number of hours spent watching T.V.	0	1	2	3	4	5	6	7
Frequency	2	6	10	5	5	3	4	5

FINDING THE MEDIAN FROM A FREQUENCY TABLE

Finding the median from a frequency table is not as straightforward as finding it from the set of numbers written out in full.

First we have to establish how many numbers are in the set. Then we have to find out which number is the median.

The following frequency table was made by asking the pupils in one class to give the number of children in their own family.

Number of children per family	1	2	3	4	5
Frequency	7	15	5	2	1

This tells us that there are 7 families with one child, 15 families with two children, 5 families with three children, ... and so on. Therefore altogether there are $7 + 15 + 5 + 2 + 1$ families, i.e. there are 30 numbers in a set.

> Adding the frequencies tells us how many numbers there are in the set.

As there are 30 numbers in the set, the median is the $\left(\frac{30+1}{2}\right)$th number.

i.e. the $15\frac{1}{2}$th number

i.e. the average of the 15th and 16th numbers.

This is easier to see if the table is written vertically.

Number of children per family	Frequency	Position in size order
1	7	1st – 7th
2	15	8th – 22nd
3	5	23rd – 27th
4	2	28th – 29th
5	1	30th
	Total 30	

Now we can see that the 15th and 16th numbers are both 2.

Therefore the median is 2.

EXERCISE 25g Find the median of the set given in each frequency table in Exercise 25f.

FINDING THE MEAN FROM A FREQUENCY TABLE

We will use the frequency table from the last section.

Number of children per family	1	2	3	4	5
Frequency	7	15	5	2	1

We already know that there are 30 numbers in this set. To find the mean we have to add all the numbers in this set and then divide by 30.

From the frequency table we can see that there are 7 ones, 15 twos, and so on,

giving a total of $(7 \times 1) + (15 \times 2) + (5 \times 3) + (2 \times 4) + (1 \times 5)$
$= 7 + 30 + 15 + 8 + 5$
$= 65$

Therefore the mean is $65 \div 30 = 2.2$ to 1 d.p.

i.e. there is an average of 2.2 children per family.

This kind of calculation needs to be done systematically and it helps if the table is written vertically. A column is then added in which we can write down the total number of children for each of the separate family sizes, i.e.

Number of children per family	Frequency	
x	f	fx
1	7	7
2	15	30
3	5	15
4	2	8
5	1	5
	Total 30	65

Therefore the mean is $\frac{65}{30} = 2.2$ to 1 d.p.

EXERCISE 25h Use the frequency tables in Exercise 25f and find the mean in each case. Give your answer correct to 1 decimal place.

GETTING INFORMATION FROM BAR CHARTS

Sometimes a set of figures is illustrated in a bar chart. Provided the axes are clearly labelled we can get information from the diagram.

A firm makes and packs screws by a fully automated process. The screws come off the production line ready packed in packets each of which contains 50 screws. To test the reliability of the production line, some of the packets were checked to see how many defective screws each one contained. The results of this test are shown in the following bar chart:

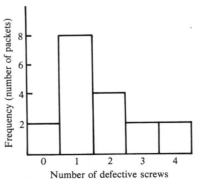

The vertical axis gives the frequency, so reading the height of the bar tells us how many packets contain no defects and how many contain one, two, three or four defective screws.

If we want to find the mode, i.e the most common number of defective screws per packet, then we can "read" this straight from the bar chart. In this case it is clearly 1.

If we want to find either the median or the mean, it is sensible first to make a frequency table from the bar chart. We had some practice in doing this in Chapter 18. In this case reading the frequencies from the top of the bars gives the following table.

No. of defective screws per packet	Frequency
x	f
0	2
1	8
2	4
3	1
4	1
	Total 16

EXERCISE 25i **1.** Some pupils were given a test which was marked out of 5. The bar chart shows the results. Find how many pupils took the test.

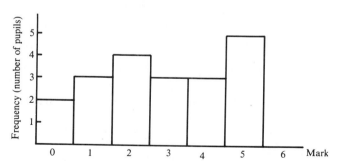

a) What is the mode?

b) Make a frequency table from the bar chart.

c) Find the median.

d) Find the mean, giving your answer correct to 1 decimal place.

Repeat parts (a) to (d) for the following bar charts.

2. The number of days that each child is away from school is recorded in a register. The bar chart illustrates the pattern of absences for a group of children for one school term. What is the total number of children in the group?

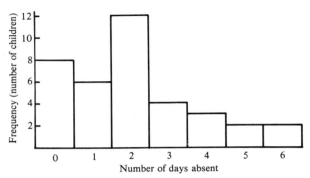

3. Apples are delivered to a Supermarket ready packed in bags, each bag containing 1 kg of apples. Some of the bags are checked to see how many apples there are in each bag. The results of this survey are illustrated in the bar chart. How many bags of apples are checked?

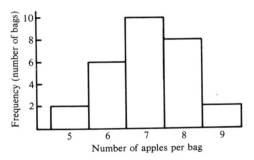

4. On one particular day a group of children were asked how many books they were taking home from school. The bar chart illustrates the results. How many children are there in the group?

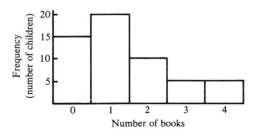

Mean, Mode and Median 345

5. A market gardener grew some cucumber plants. He counted how many cucumbers he got from each plant and the results are illustrated in the bar chart. How many plants did he grow?

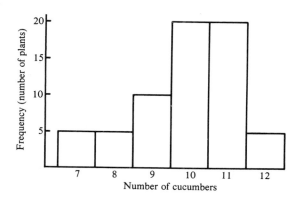

EXERCISE 25j In this exercise each question is followed by several alternative answers. Write down the letter that corresponds to the correct answer.

1. The median of 2, 5, 6, 7, 2 is

    A 2   **B** 6   **C** 5   **D** 7

2. The mode of 4, 2, 3, 4, 5 is

    A 4   **B** 2   **C** 5   **D** 3

3. The mean of 2, 2, 3, 5 is

    A $2\frac{1}{2}$ **B** 5   **C** 4   **D** 3

4. The bar chart shows the test marks for 15 pupils.

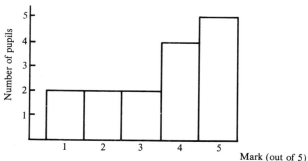

The median mark is

    A 3   **B** 4   **C** $3\frac{1}{2}$

5. The frequency table shows the number of bad apples in each of 10 boxes of apples.

Number of bad apples in a box	0	1	2
Frequency	5	3	2

The mean number of bad apples per box is

A 0 **B** 7 **C** $\frac{1}{2}$ **D** $\frac{7}{10}$

6. The bar chart shows the marks scored by a group of pupils in a test.

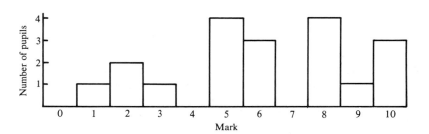

The number of pupils scoring 4 or more is

A 4 **B** 15 **C** 13 **D** 19

7.

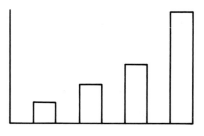

Put your money in

BLACK HOLE
UNIT TRUST

for value and security

The diagram shows

A that the number of customers is increasing every year.
B that the capital is increasing
C that the dividends are getting larger each year
D nothing.

8.

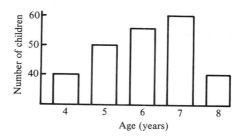

The diagram shows the results of a survey on the ages of children using a playground in a park.

Which of the following statements must be true?

A Twice as many five year olds as four year olds used the playground.

B The number of seven year olds was three times the number of eight year olds.

C More seven year olds used the playground than any other age group.

D Young children do not like the playground.

9.

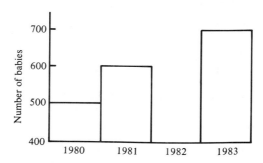

The diagram shows the results of a survey on the number of babies born in different years in Toytown.

Which of the following statements must be true?

A No children were born in 1982.

B The birth rate trebled from 1980 to 1983.

C 100 more children were born in 1981 than in 1980.

D Twice as many children were born in 1981 as in 1980.

26 PYTHAGORAS' THEOREM

SQUARES

We obtain the square of a number when we multiply the number by itself, e.g. the square of 5 is 5×5 which is 25

We write $\quad 5^2 = 25$

Similarly $13^2 = 169$ and $2.4^2 = 5.76$

EXERCISE 26a

> Find 0.07^2
>
> $$0.07^2 = \underset{(2dp)}{0.07} \times \underset{(2dp)}{0.07}$$
>
> $$= \underset{(4dp)}{0.0049}$$

Without using a calculator, write down the squares of the following numbers.

1.	4	**5.**	3	**9.**	6
2.	8	**6.**	7	**10.**	10
3.	12	**7.**	9	**11.**	20
4.	30	**8.**	40	**12.**	70
13.	0.3	**17.**	0.4	**21.**	0.6
14.	1.1	**18.**	1.2	**22.**	0.2
15.	0.7	**19.**	0.5	**23.**	0.07
16.	0.02	**20.**	0.06	**24.**	0.004

> Use a calculator to find the square of a) 4.7 b) 12.3 c) 64.8. Give your answers correct to 4 s.f.
>
> a) $4.7^2 = 22.09$ b) $12.3^2 = 151.3$ c) $64.8^2 = 4199$

Pythagoras' Theorem

Use a calculator to find the squares of the following numbers. Give your answers correct to 4 s.f.

25.	8.7	**29.**	7.3	**33.**	9.4
26.	4.3	**30.**	5.6	**34.**	6.5
27.	12.2	**31.**	24.5	**35.**	17.9
28.	33.6	**32.**	47.8	**36.**	23.7
37.	1.732	**41.**	1.414	**45.**	2.236
38.	37.32	**42.**	29.71	**46.**	56.24
39.	8.734	**43.**	6.554	**47.**	3.928
40.	69.45	**44.**	78.22	**48.**	83.17

SQUARE ROOTS

The square root of a number is the number which when multiplied by itself, gives the original number,

e.g. since $5^2 = 25$, the square root of 25 is 5.

The square root could also be -5, since $(-5) \times (-5) = 25$ but here we are concerned only with positive square roots.

We write $\sqrt{25} = 5$

EXERCISE 26b Write down the square root of the following numbers, without using a calculator.

1.	16	**5.**	4	**9.**	36
2.	81	**6.**	64	**10.**	49
3.	1	**7.**	100	**11.**	144
4.	9	**8.**	121	**12.**	25
13.	0.25	**17.**	0.16	**21.**	0.36
14.	0.09	**18.**	0.04	**22.**	0.01
15.	0.49	**19.**	1.44	**23.**	0.81
16.	0.0036	**20.**	0.0004	**24.**	1.21

> Use a calculator to find, correct to 4 s.f., the square root of
> a) 39.2   b) 527   c) 183.4
>
> Check your answers by squaring them.
>
> a) $\sqrt{39.2} = 6.261$ $(6.261^2 = 39.20)$
> b) $\sqrt{527} = 22.96$ $(22.96^2 = 527.2)$
> c) $\sqrt{183.4} = 13.54$ $(13.54^2 = 183.4)$

Use a calculator to find, correct to 4 s.f., the square roots of the following numbers. Check your answers by squaring them.

25. 84
26. 147
27. 727
28. 19
29. 39
30. 432
31. 29
32. 138
33. 76

34. 62.3
35. 519.7
36. 3.926
37. 54.29
38. 42.7
39. 224.8
40. 5.443
41. 73.16
42. 73.8

EXERCISE 26c Use a calculator to find the following squares or square roots. Give your answers correct to 4 s.f. where necessary. Check your answers using the reverse process.

1. 5.47^2
2. $\sqrt{34.2}$
3. 75.3^2
4. $\sqrt{240}$
5. 3.64^2
6. $\sqrt{84.3}$
7. 12.6^2
8. $\sqrt{500}$
9. 1.993^2
10. $\sqrt{50.2}$
11. 57.7^2
12. $\sqrt{747}$

13. $\sqrt{1.747}$
14. 1.747^2
15. 79.24^2
16. $\sqrt{349.2}$
17. $\sqrt{9.345}$
18. 9.345^2
19. 18.91^2
20. $\sqrt{763.8}$
21. $\sqrt{29.21}$
22. 29.21^2
23. 57.67^2
24. $\sqrt{693.4}$

PYTHAGORAS' THEOREM

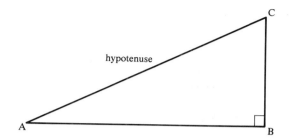

Pythagoras' Theorem states that in a right-angled triangle the square of the hypotenuse is equal to the sum of the squares of the other two sides,

i.e. in the given triangle

$$AC^2 = AB^2 + BC^2$$

EXERCISE 26d Write Pythagoras' result for each of the following figures. In some questions part of the result is already given.

1.

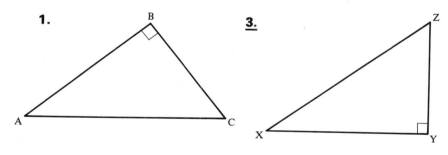

3.

2.

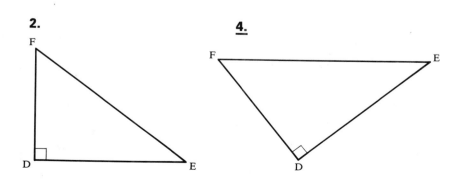

4.

In questions 5 to 10 use Pythagoras' result to complete the statements.

5.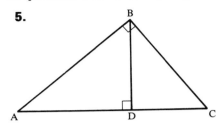

a) $AB^2 =$

b) $\quad\quad = BD^2 + DC^2$

8.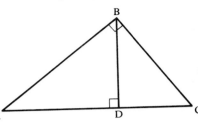

a) $AC^2 =$

b) $\quad\quad = AD^2 + BD^2$

c) $\quad\quad = CD^2 +$

6.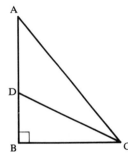

a) $AC^2 =$

b) $\quad\quad = BD^2 +$

9.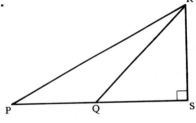

a) $\quad\quad = QS^2 + RS^2$

b) $\quad\quad = PS^2 + RS^2$

7.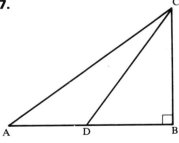

a) $DB^2 + BC^2 =$

b) $AC^2 - BC^2 =$

c) $CB^2 = CD^2 -$

and $CB^2 = AC^2 -$

10.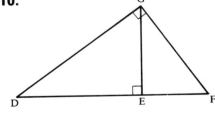

a) $GE^2 = \quad\quad - EF^2$

b) $DE^2 + EG^2 =$

c) $GF^2 = DF^2 -$

If we are given any two sides of a right-angled triangle we can find the third side.

EXERCISE 26e In this exercise, when taking values from your calculator, write down 4 significant figures. Give your answer correct to 3 s.f.

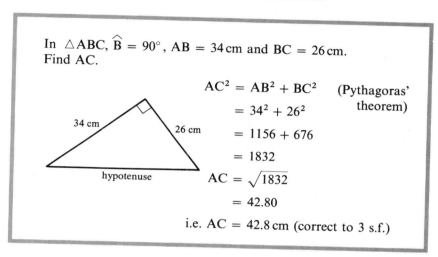

In $\triangle ABC$, $\widehat{B} = 90°$, $AB = 34$ cm and $BC = 26$ cm. Find AC.

$$AC^2 = AB^2 + BC^2 \quad \text{(Pythagoras' theorem)}$$
$$= 34^2 + 26^2$$
$$= 1156 + 676$$
$$= 1832$$
$$AC = \sqrt{1832}$$
$$= 42.80$$

i.e. AC = 42.8 cm (correct to 3 s.f.)

Use the information given in the diagrams to find the required lengths.

1. Find BC.

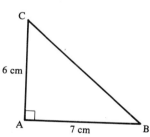

3. Find PR.

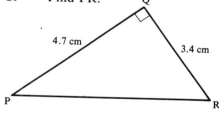

2. Find DF.

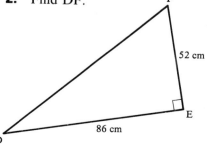

4. Find LM.

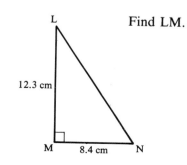

5. Find AC.

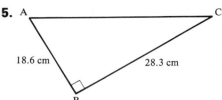

8. Find AB.

6. Find XY.

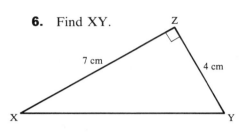

9. Find BC.

7. Find PR.

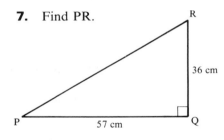

10. Find DF.

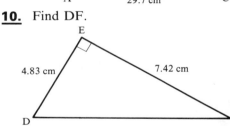

In ABC, $\widehat{B} = 90°$, AC = 4.16 cm and AB = 3.27 cm.
Find BC.

$$AC^2 = AB^2 + BC^2 \quad \text{(Pythagoras' result)}$$
$$4.16^2 = 3.27^2 + BC^2$$
$$17.31 = 10.69 + BC^2$$

Take 10.69 from each side
$$6.62 = BC^2$$
$$BC = \sqrt{6.62}$$
$$= 2.573$$

i.e. BC = 2.57 cm (correct to 3 s.f.)

11. 12 cm, 7 cm. Find AB.

12. 9 cm, 21 cm. Find LN.

13. 19 cm, 34 cm. Find PQ.

14. 10.9 cm, 8.7 cm. Find XZ.

15. 13.41 cm, 8.37 cm. Find AB.

16. 9 cm, 13 cm. Find PR.

17. 33 cm, 18 cm. Find XY.

18. 52 cm, 26 cm. Find AB.

19. 6.9 cm, 12.3 cm. Find EF.

20. 24.6 cm, 18.7 cm. Find PR.

TWO SPECIAL RIGHT-ANGLED TRIANGLES

You will have noticed that in most cases when two sides of a right-angled triangle are given and the third side is calculated using Pythagoras' Theorem, the answer is not an exact number. There are, however a few special cases where all three sides are exact numbers.

The simplest one is the triangle with sides of 3, 4 and 5 units.

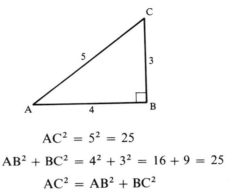

$$AC^2 = 5^2 = 25$$
$$AB^2 + BC^2 = 4^2 + 3^2 = 16 + 9 = 25$$

i.e. $$AC^2 = AB^2 + BC^2$$

Any triangle similar to this one has sides in the ratio 3 : 4 : 5 so whenever you spot this case you can find the missing side very easily.

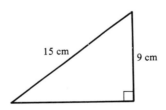

For instance, in this triangle

$$15 = 5 \times 3, \text{ and } 9 = 3 \times 3.$$

The triangle is similar to the 3, 4, 5 triangle, so the remaining side is 4×3 cm, i.e., 12 cm.

The other frequently used right-angled triangle with exact sides is the 5, 12, 13 triangle.

$$13^2 = 169$$

and $$5^2 + 12^2 = 25 + 144 = 169$$

so $$13^2 = 5^2 + 12^2$$

EXERCISE 26f

In △ABC, B = 90°, AB = 24 cm and BC = 18 cm. Find AC.

BC = 6 × 3 cm

and AB = 6 × 4 cm

so AC = 6 × 5 cm   (3, 4, 5 △)

= 30 cm

In each of the following questions, decide whether the triangle is similar to the 3, 4, 5 triangle or to the 5, 12, 13 triangle, and hence find the missing side.

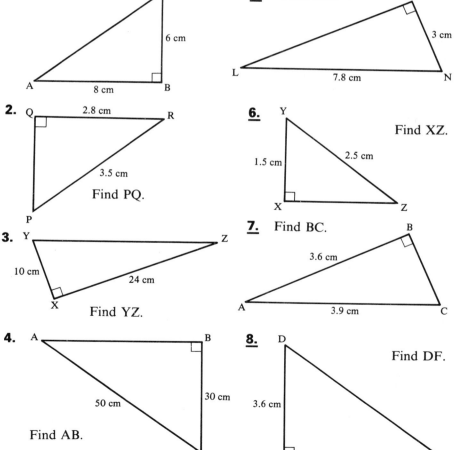

1. Find AC.

2. Find PQ.

3. Find YZ.

4. Find AB.

5. Find LM.

6. Find XZ.

7. Find BC.

8. Find DF.

GRAPHICAL USE OF PYTHAGORAS' THEOREM TO FIND THE SQUARE ROOTS OF THE NATURAL NUMBERS

Start with a right-angled triangle and find the hypotenuse when the other two sides are of unit length (1 inch is a good unit to use).

Use each hypotenuse, together with a side of 1 unit as the sides forming the right angle for the next triangle. You will find that the answers get more inaccurate as you go up but $\sqrt{4} = 2$ and $\sqrt{9} = 3$ will act as checks on the accuracy of your work.

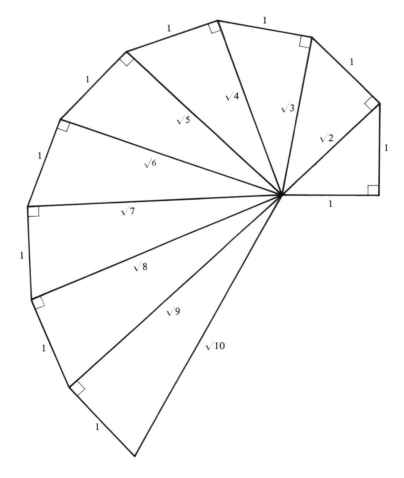

Use this method to find the square root of each whole number up to 10.
Compare these results with the values obtained from your calculator. If your values are accurate you might like to continue as far as 20.

PROBLEMS

When using a calculator, write your intermediate calculations to four significant figures. In this exercise give all your answers correct to three significant figures.

EXERCISE 26g

A wire stay supporting a telegraph pole is 10 m long. It is attached to a point 1 m from the top of the pole and to a point 4 m from the base of the pole on level ground. Find the height of the telegraph pole.

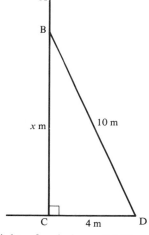

Let ABC represent the vertical pole and BD the wire stay.
Let the distance BC be x m.

In triangle BCD Pythagoras' result gives

$$BD^2 = x^2 + CD^2$$
i.e. $10^2 = x^2 + 4^2$
$$100 = x^2 + 16$$
$$84 = x^2$$
$$\therefore x = 9.165$$

Height of pole is $\quad 9.165\,\text{m} + 1\,\text{m} = 10.165\,\text{m}$

$\qquad\qquad\qquad\qquad\quad = 10.2\,\text{m}$ correct to 3 s.f.

Remember to draw a diagram for each question.

1. A football pitch measures 120 yd by 80 yd. How far is it between opposite corners?

2. A hockey pitch measures 60 m by 95 m. Find the length of a diagonal of the pitch.

3. A rugby pitch measures 100 m by 70 m. How far is it between opposite corners?

4. A tennis court measures 29.6 m by 10.7 m. How far is it between opposite corners? Give your answer correct to the nearest tenth of a metre.

5. One diagonal of a front door, which is 0.813 m wide, is 2.03 m. How high is the door?

6. The diagonal of a rectangular sheet of glass is 73 cm. If the rectangle is 58 cm long, how wide is it?

Questions 7–15 refer to the chess board given below.

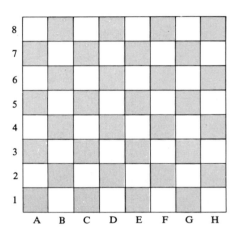

Assume that the side of each square is 4 cm.

How far is it from

7. the bottom left corner of A1 to the top right corner of H8?

8. the bottom left corner of A1 to the top right corner of G4?

9. the bottom left corner of C3 to the top right corner of F8?

10. the top left corner of B2 to the bottom right corner of E7?

11. the top right corner of B3 to the bottom right corner of H6?

12. the top right corner of D2 to the top left corner of G5?

13. the top right corner of A1 to the top left corner of E8?

14. the bottom right corner of B1 to the top left corner of G7?

15. the bottom left corner of C2 to the bottom left corner of H8?

16. A wire stay, 12 m long, is attached to a flag pole 2 m from the top. The other end of the wire is fixed to a point in level ground 9 m from the base of the pole. How high is the pole?

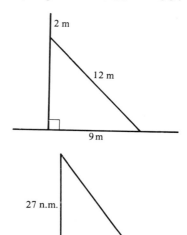

17. A ship sails 27 nautical miles due south, then 20 nautical miles due east. How far is it from its starting point?

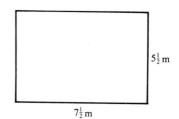

18. A rectangular classroom measures $7\frac{1}{2}$ m by $5\frac{1}{2}$ m. What is the length of the longest straight line that can be drawn on the floor?

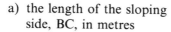

19. The diagram shows the sections of the ridged roof of a farm building. The section is 8 m wide and $1\frac{1}{2}$ m high. If the building is 12 m long find

a) the length of the sloping side, BC, in metres

b) the total area of the sloping roof in square metres.

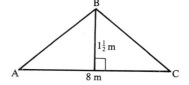

20. The rear doors of a van each measure 2 m by $\frac{3}{4}$ m. Find the width of the largest board of negligible thickness that can be loaded into the van.

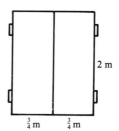

EXERCISE 26h In this exercise several alternative answers are given. For each question write down the letter that corresponds to the correct answer.

1. The square of 0.4 is

   A 0.016 **B** 1.6 **C** 0.16 **D** 0.2

2. The square root of 0.0009 is

   A 0.003 **B** 0.03 **C** 0.0003 **D** 0.81

3.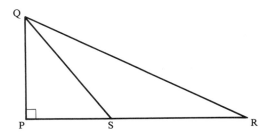

   In the above diagram PQ^2 is equal to

   A $QS^2 + PS^2$ **C** $QR^2 - PR^2$
   B $QR^2 - SR^2$ **D** none of these

4.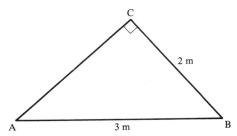

   Using the information given in the diagram, the length of AC correct to 1 d.p. is

   A 2.2 m **B** 1 m **C** 3.6 m **D** 2 m

5. The diagonal of a rectangular cupboard door is 50 cm. If the door is 40 cm high its width is

   A 41 cm **B** 40 cm **C** 30 cm **D** 64 cm

EXERCISE 26i

1. Use a calculator to find, correct to 3 s.f.   a) 0.874^2 b) 5.93^2

2. Use a calculator to find, correct to 3 s.f.   a) $\sqrt{0.0874}$ b) $\sqrt{593}$

3. Use the information given in the diagram to find QR.

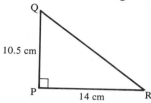

4. Use the information given in the diagram to find BC.

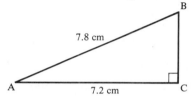

5. The diagonals of a rhombus have lengths 12 cm and 16 cm. Find the length of a side.

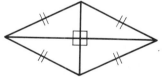

6. Use the information given in the diagram to find

   a) BC   b) AC   c) AD

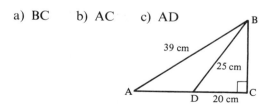

7. Calculate, correct to 3 s.f., the diagonal of a square of side 15 cm.

27 AREAS

REVISION OF BASIC UNITS

Units of length The basic unit of length is the metre (m).
Other units in common use are the centimetre (cm), the millimetre (mm) and the kilometre (km), where

$$1 \text{ km} = 1000 \text{ m}$$
$$1 \text{ m} = 100 \text{ cm}$$
$$1 \text{ cm} = 10 \text{ mm}$$

⊢——⊣ 1 cm

Units of area These are derived from the units of length and those in common use are the square metre (m^2), the square centimetre (cm^2), the square millimetre (mm^2) and the hectare (h), where

$$1 \text{ h} = 1000 \text{ m}^2$$
$$1 \text{ m}^2 = 100 \times 100 \text{ cm}^2 = 10000 \text{ cm}^2$$
$$1 \text{ cm}^2 = 10 \times 10 \text{ mm}^2 = 100 \text{ mm}^2$$

□ 1 cm^2

EXERCISE 27a

> Express 2.5 m in centimetres.
>
> $$2.5 \text{ m} = 2.5 \times 100 \text{ cm}$$
> $$= 250 \text{ cm}$$

Express the given quantity in the unit in brackets.

1. 5.3 km (m)   **4.** 52 cm (mm)   **7.** 18 cm (mm)
2. 2.8 cm (mm)   **5.** 5 m (mm)   **8.** 2 m (mm)
3. 1.7 m (cm)   **6.** 2.3 m (cm)   **9.** 8.7 km (m)

> Express 57 cm in metres.
>
> $$57 \text{ cm} = \frac{57}{100} \text{ m}$$
> $$= 0.57 \text{ m}$$

Express the given quantity in the unit in brackets.

10. 250 mm (cm) **15.** 872 mm (cm)
11. 50 cm (m) **16.** 2850 m (km)
12. 1000 m (km) **17.** 28 mm (cm)
13. 350 mm (cm) **18.** 180 m (km)
14. 2500 m (km) **19.** 650 cm (m)

Express $2\,m^2$ in square centimetres.

$$2\,m^2 = 2 \times 100 \times 100\,cm^2$$
$$= 20\,000\,cm^2$$

Express the given quantity in the unit in brackets.

20. $5\,cm^2$ (mm^2) **24.** $3\,cm^2$ (mm^2)
21. 2 h (m^2) **25.** $0.5\,m^2$ (cm^2)
22. $0.2\,m^2$ (cm^2) **26.** $2.6\,cm^2$ (mm^2)
23. $0.5\,cm^2$ (mm^2) **27.** 5.5 h (m^2)

Express $2500\,m^2$ in hectares.

$$2500\,m^2 = \frac{2500}{1000}\,h$$
$$= 2.5\,h$$

Express the given quantity in the unit in brackets.

28. $300\,mm^2$ (cm^2) **32.** $560\,cm^2$ (m^2)
29. $5000\,m^2$ (h) **33.** $5680\,mm^2$ (cm^2)
30. $2500\,cm^2$ (m^2) **34.** $4080\,m^2$ (h)
31. $27\,mm^2$ (cm^2) **35.** $20\,000\,cm^2$ (m^2)

36. 5000 m (km)
37. 2.7 cm (mm)
38. 36 cm² (mm²)
39. 2500 m² (h)

40. 650 cm (m)
41. 8000 m² (h)
42. 2.5 km² (m²)
43. 3.8 cm (mm)

44. 5.3 mm (cm)
45. 850 cm² (mm²)
46. 380 cm (m)
47. 9.8 m (cm)

48. 0.5 cm (mm)
49. 500 mm² (cm²)
50. 82 cm (m)
51. 9000 m² (h)

PERIMETERS

The perimeter of a plane figure is the distance all round it, i.e. the sum of the lengths of its sides.

EXERCISE 27b

Find the perimeter of the given figure.

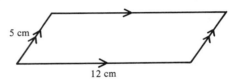

(This is a parallelogram, so the opposite sides are equal in length.)

Perimeter = (5 + 12 + 5 + 12) cm

= 34 cm

Find the perimeter of each of the following figures.

1.

5 m
5 m

2.

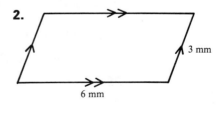

3 mm
6 mm

Areas 367

3.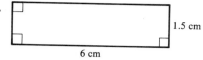

8.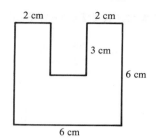

4.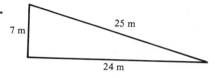

9.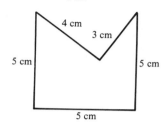

5.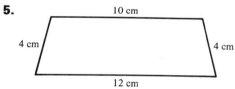

10.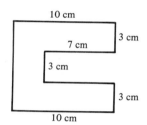

6.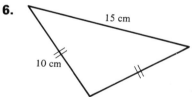

11.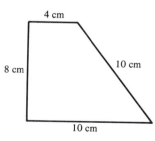

7.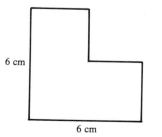

12.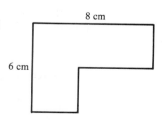

AREA OF A RECTANGLE

The area of a rectangle is found by multiplying its length by its breadth.

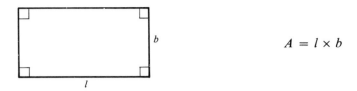

$A = l \times b$

A square is a rectangle whose length and breadth are the same.

$A = l \times l$
$= l^2$

EXERCISE 27c Remember that area is measured in square units.

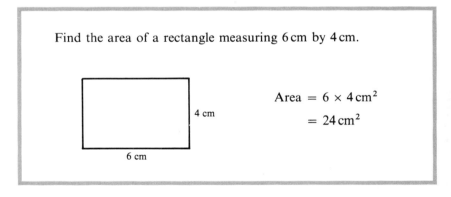

Find the area of a rectangle measuring 6 cm by 4 cm.

Area $= 6 \times 4 \, \text{cm}^2$
$= 24 \, \text{cm}^2$

Find the area of each of the following shapes. Remember that you must state the units involved.

1. A square of side 8 m.

2. A rectangle measuring 4 cm by 3 cm.

3. A square of side 2 mm.

4. A rectangle measuring 4 m by 8 m.

5. A rectangle measuring 10 m by 12 m.

Areas 369

6. A square of side 5 cm.
7. A rectangle of length 20 mm and breadth 8 mm.
8. A rectangle measuring 1.5 cm by 4 cm.
9. A square of side 2.5 m.
10. A rectangle measuring 1.2 cm by 2.3 cm.

Remember that, when we find an area, the measurements that we use must be in the *same* unit. We usually change the larger unit to the smaller unit.

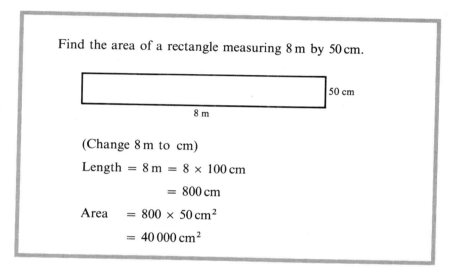

Find the area of a rectangle measuring 8 m by 50 cm.

(Change 8 m to cm)

Length = 8 m = 8 × 100 cm
= 800 cm
Area = 800 × 50 cm²
= 40 000 cm²

Find the areas of the following rectangles.

11. Length 1 cm, breadth 5 mm.
12. Length 1 m, breadth 20 cm.
13. Length 0.5 cm, breadth 3 mm.
14. Length 1 m, breadth 30 cm.
15. Length 2 cm, breadth 4 mm.
16. Length 3 m, breadth 70 cm.
17. Length 10 cm, breadth 9 mm.
18. Length 0.3 m, breadth 22 cm.
19. Length 100 mm, breadth 0.4 cm.
20. Length 200 cm, breadth 0.6 m.

AREA OF A PARALLELOGRAM

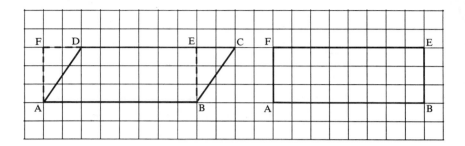

If we cut △BCE from parallelogram ABCD and move it to the other end of the parallelogram as shown, we get the rectangle ABEF.

Therefore the area of parallelogram ABCD is equal to the area of the rectangle ABEF.

Therefore area of parallelogram = AB × BE

= base × perpendicular height

When we use the word *height* we mean the *perpendicular height*, not the slant height, therefore

> Area of parallelogram = base × height

EXERCISE 27d

Find the area of the given parallelogram.

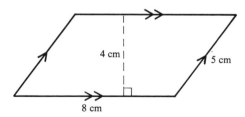

Area = base × perpendicular height

= 8 × 4 cm²

= 32 cm²

(Notice that we do not use the length of the 5 cm side.)

Areas 371

Find the areas of the following parallelograms.

1.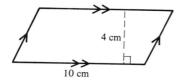

6.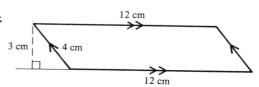

2.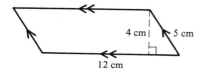

7.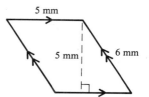

3.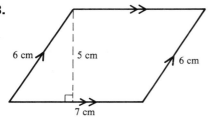

8.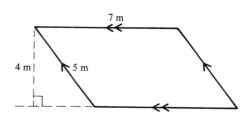

4.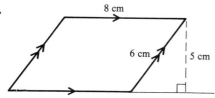

9.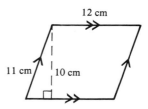

5.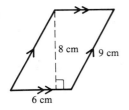

10.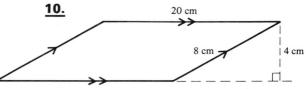

In questions 11 to 16 turn the page round if necessary so that you can see which is the base and which the perpendicular height.

11.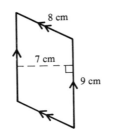

14.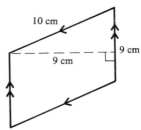

12.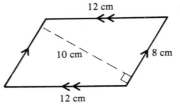

15.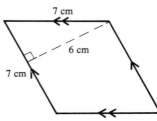

13.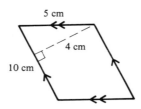

16.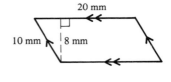

THE AREA OF A TRIANGLE

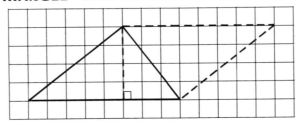

We can think of a triangle as half of a parallelogram, therefore area of triangle = $\frac{1}{2}$ (area of parallelogram)

i.e. | Area of triangle = $\frac{1}{2}$ × base × perpendicular height |

As is the case with a parallelogram, when we say 'the height of a triangle' we mean the perpendicular height.

Areas 373

EXERCISE 27e

Find the area of the triangle in the diagram.

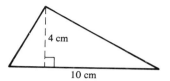

Area = $\frac{1}{2}$ × base × perpendicular height

   = $\frac{1}{2}$ × 10 × 4 cm²

   = 20 cm²

Find the areas of the following triangles.

1.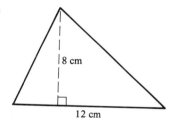

2.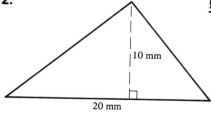

3.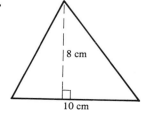

4.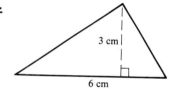

5.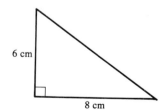

6.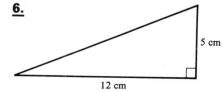

7.
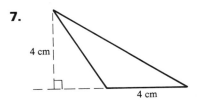

9.
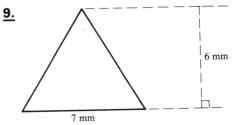

8.
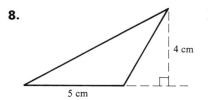

10.
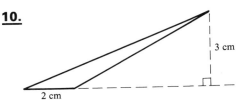

In numbers 11 to 16 turn the page round if necessary so that you can see which is the base and which is the perpendicular height.

11.
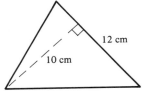

14.
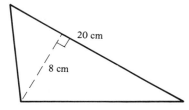

12.
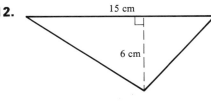

15.
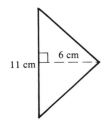

13.
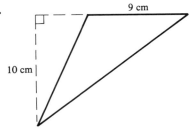

16.
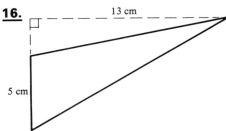

THE AREA OF A TRAPEZIUM

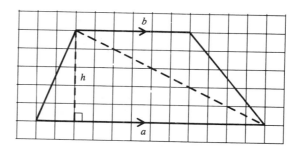

By drawing one diagonal we can divide a trapezium into two triangles whose heights are the same.

Therefore the area of the trapezium is equal to the sum of the areas of the two triangles

i.e. $\qquad$ area of trapezium $= \frac{1}{2}ah + \frac{1}{2}bh$

Now $\frac{1}{2}$ and h are common factors of this expression, so

$$\text{area of trapezium} = \frac{1}{2}(a+b) \times h$$

This formula is easier to remember if it is given in words.

> The area of a trapezium is equal to
> $\frac{1}{2}$(sum of parallel sides) × (perpendicular distance between them)

EXERCISE 27f

Find the area of the trapezium in the diagram.

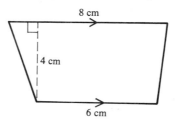

$$\begin{aligned}
\text{Area} &= \tfrac{1}{2}(\text{sum of parallel sides}) \times (\text{distance between them}) \\
&= \tfrac{1}{2}(8+6) \times 4 \, \text{cm}^2 \\
&= \tfrac{1}{2} \times 14 \times 4 \, \text{cm}^2 \\
&= 28 \, \text{cm}^2
\end{aligned}$$

Find the areas of the following trapeziums.

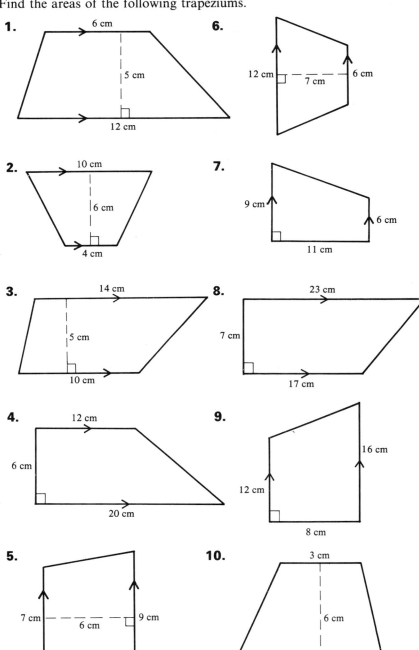

MIXED QUESTIONS

Remember that 'height' means 'perpendicular height' and that the 'distance' between two lines means 'perpendicular distance'.

EXERCISE 27g The shapes in this exercise are drawn on squared paper. Each square is of side 0.5 cm. Copy each figure on to squared paper and for each question

a) name the shape.

b) on your diagram mark the lengths (in cm) of the lines you need to find its area. (You may have to draw an extra line for the height or for the distance between parallel sides.)

c) find the area of the shape in cm².

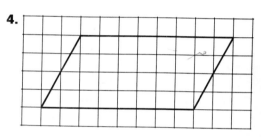

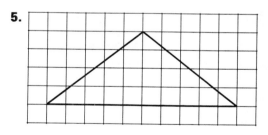

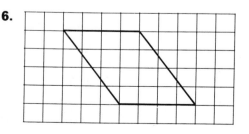

7.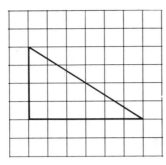

11.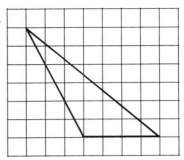

8.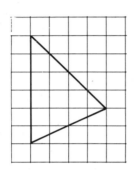

12.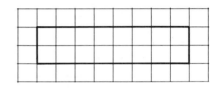

9.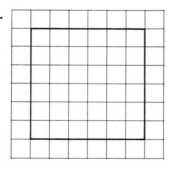

13.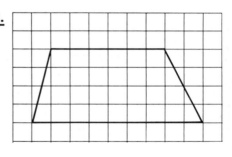

10.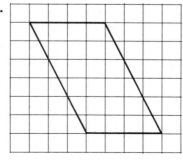

14.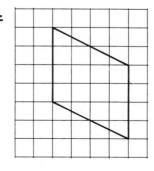

Areas 379

15.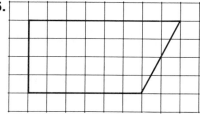

19.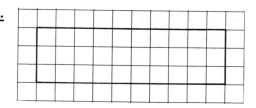

16.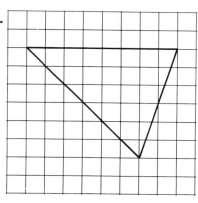

20.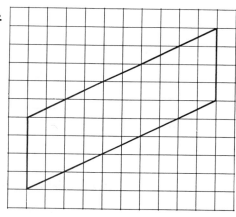

17.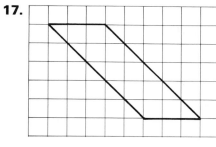

21.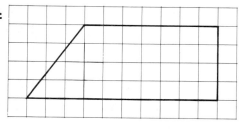

18.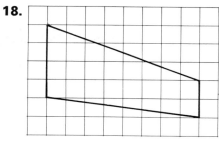

22.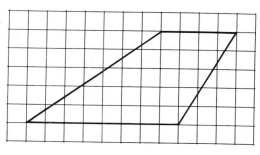

AREAS OF COMPOUND SHAPES

It is often possible to find the area of a figure by dividing it into two or more shapes whose areas can be found.

EXERCISE 27h

Find the area of the figure in the diagram.

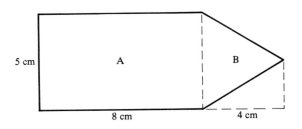

(A is a rectangle, 8 cm by 5 cm
B is a triangle, base 5 cm, height 4 cm.)

Area of A = 8×5 cm²
= 40 cm²

Area of B = $\frac{1}{2}$ × base × perpendicular height
= $\frac{1}{2} \times 5 \times 4$ cm²
= 10 cm²

therefore the total area = 50 cm²

In questions 1 to 8, each figure is divided into two shapes A and B. In each case

a) copy the figure.

b) name the shapes A and B and mark any extra measurements needed to find the areas.

c) find the area of A.

d) find the area of B.

e) find the total area of the Figure.

Areas 381

1.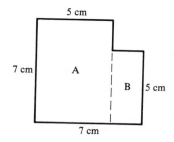

2.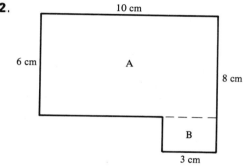

3.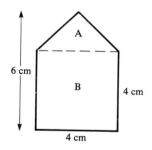

4.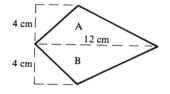

5.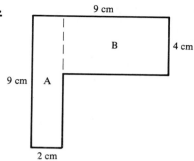

6.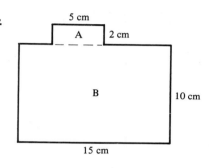

7.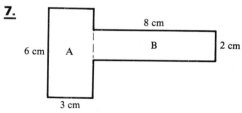

8.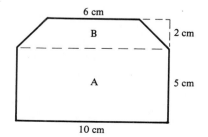

For the given figure find

a) the perimeter

b) the area.

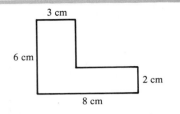

(From the measurements given we can work out the lengths of the other sides.)

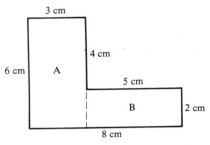

a) Perimeter $= (6 + 3 + 4 + 5 + 2 + 8)$ cm

$= 28$ cm

b) Area of A $= 6 \times 3$ cm^2

$= 18$ cm^2

Area of B $= 5 \times 2$ cm^2

$= 10$ cm^2

Total area $= (18 + 10)$ cm$^2 = 28$ cm^2

In each of the following questions find

a) the perimeter   b) the area   of the given figure.

9.

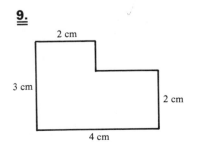

10.

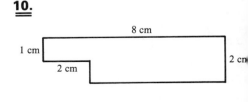

Areas 383

11.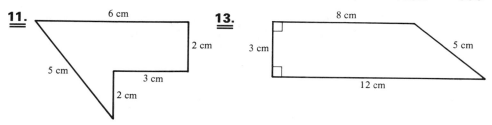

13.

12.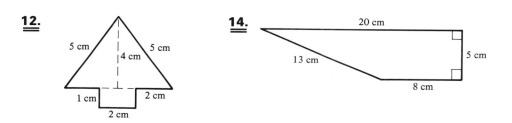

14.

PROBLEMS

EXERCISE 27i **1.** The diagram represents a rectangular sports ground. The shaded area is a rectangular pitch.

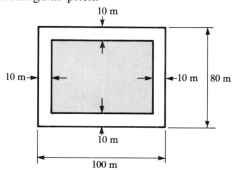

For the pitch, find a) the length
 b) the width
 c) the perimeter
 d) the area.

2. The diagram shows the layout of a rectangular garden. The shaded area is a paved patio, the rest is grass.

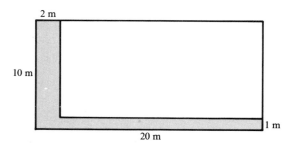

Find a) the perimeter of the whole garden

b) the length of the lawn

c) the width of the lawn

d) the area of the lawn

e) the area of the patio

f) the perimeter of the lawn.

3. The diagram shows the end wall of a garden shed. The shaded area is a door.

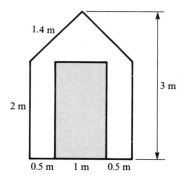

Find a) the width of the wall

b) the area of the wall, including the door

c) the area of the door

d) the area of the unshaded part of the wall.

4. The diagram represents a rectangular allotment.
The shaded area is path, the rest is a vegetable plot.

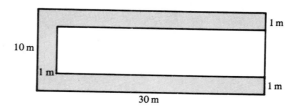

Find a) the length of the vegetable plot
b) the width of the vegetable plot
c) the area of the vegetable plot
d) the area of the whole allotment
e) the area of the path
f) the perimeter of the vegetable plot
g) the perimeter of the shaded area.

5. The diagram shows the floor plan of a room.
The shaded area is carpet, the rest is polished wood.

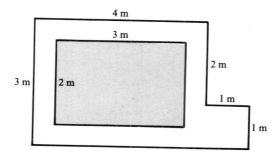

Find the area   a) of the carpet
b) of the whole floor
c) of the uncarpeted part of the floor.

6. The diagram shows the floor plan of a room with a bay.

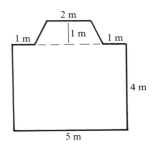

Find   a) the area of the rectangular part of the floor

b) the area of the bay

c) the total floor area of the room.

EXERCISE 27j In these questions you are given several alternative answers. Write down the letter that corresponds to the correct answer.

1.

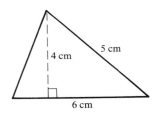

The area of the triangle is

A 30 m²   **B** 24 cm²   **C** 15 cm²   **D** 12 cm²

2.

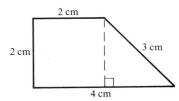

The perimeter of the figure is

A 8 cm   **B** 11 cm   **C** 6 cm²   **D** 11 cm²

3.

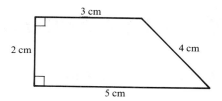

The area of the figure is

A 16 cm²   **B** 8 cm²   **C** 14 cm²   **D** 10 cm²

4.

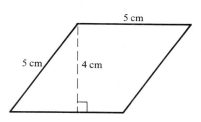

The area of the figure is

A 25 cm²   **B** 10 cm²   **C** 9 cm²   **D** 20 cm²

5.

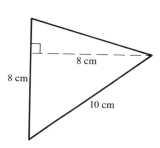

The area of the figure is

A 16 cm²   **B** 32 cm²   **C** 40 cm²   **D** 64 cm²

28 CIRCLE CALCULATIONS

DIAMETER, RADIUS AND CIRCUMFERENCE

The following diagram gives a reminder of the names of the parts of a circle.

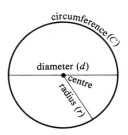

A diameter goes through the centre of a circle (i.e. across the full width) so the diameter is twice as long as the radius.

i.e. $$d = 2r$$

EXERCISE 28a Write down the length of the diameter of each of the following circles.

1.

2 mm

4.

2.5 cm

2.
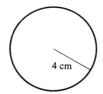
4 cm

5.

3 mm

3.

1.5 cm

6.
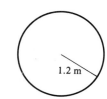
1.2 m

Write down the length of the radius of each of the following circles.

7.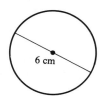

9.

8.

10.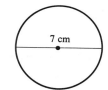

THE CIRCUMFERENCE OF A CIRCLE

In book 2 we saw that, whatever the size of a circle, the circumference is about three times the length of the diameter.

To get the circumference we have to multiply the diameter by a number that is a little bigger than 3. We cannot write this number exactly using figures so we call it π.

Therefore   circumference = π × diameter

or $C = \pi d$

or $C = 2\pi r$

where r is the radius.

To as many figures as we can write across the page

π = 3.14159265358979323846264338327950288419716939937510582097

We never need this degree of accuracy, so we usually take π as 3.14 or 3.142, or the value stored in some calculators. You will be told what value to use for π.

EXERCISE 28b For all questions use your calculator, take π as 3.142 and give your answers correct to three significant figures.

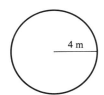

Find the circumference of a circle of radius 4 m

$$C = 2\pi r$$
$$= 2 \times 3.142 \times 4 \, m$$
$$= 25.13 \, m$$
$$= 25.1 \, m \text{ (correct to 3 s.f.)}$$

Find the circumference of a circle of radius

1. 6 cm **4.** 3 m **7.** 10 cm
2. 2 cm **5.** 2 mm **8.** 30 m
3. 4 cm **6.** 8 m **9.** 12 m

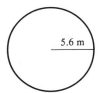

Find the circumference of a circle of radius 5.6 cm

$$C = 2\pi r$$
$$= 2 \times 3.142 \times 5.6 \, cm$$
$$= 35.19 \, cm$$
$$= 35.2 \, cm \text{ (correct to 3 s.f.)}$$

Find the circumference of a circle of radius

10. 38 mm **13.** 3.5 cm **16.** 12.6 cm
11. 4.4 m **14.** 2.6 cm **17.** 9.3 cm
12. 8.2 m **15.** 10.5 cm **18.** 315 cm

Find a) the radius b) the circumference of the circle whose *diameter* is given.

19. 56 cm **22.** 14 cm **25.** 6.3 cm
20. 8.4 cm **23.** 2.8 cm **26.** 2.1 cm
21. 98 cm **24.** 3.5 cm **27.** 42 cm

Circle Calculations 391

SECTORS AND ARCS

Part of the circumference of a circle is called an *arc*.
A "slice" of a circle is called a *sector*.

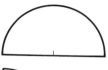

Half a circle is called a *semi-circle*.

One quarter of a circle is called a *quadrant*.

EXERCISE 28c

A shape is made of a square of side 2 m with a quadrant of a circle on one side.

Use $\pi = 3.142$ to find the perimeter of the shape.

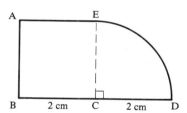

(The lengths of the three straight sides can be found from the diagram because ABCE is a square. We will have to calculate the length of the arc ED.)

The arc ED is one quarter of the circumference of a circle of radius 2 cm.

Therefore the length of arc ED $= \frac{1}{4}(2\pi r)$

$\qquad\qquad\qquad\qquad\qquad = \frac{1}{4} \times 2 \times 3.142 \times 2$ cm

$\qquad\qquad\qquad\qquad\qquad = 3.142$ cm

Therefore the perimeter $=$ arc ED $+$ BD $+$ AB $+$ AE

$\qquad\qquad\qquad\qquad = (3.142 + 4 + 2 + 2)$ cm

$\qquad\qquad\qquad\qquad = 11.1$ cm (correct to 3 s.f.)

Use π = 3.142 for this exercise.

1.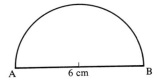

The figure is a semicircle. Find
a) the radius of the semicircle
b) the length of the arc AB
c) the perimeter of the figure.

2.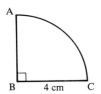

The figure is a quadrant of a circle. Find
a) the length of the arc AC
b) the perimeter of the figure.

3.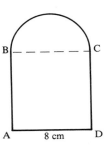

The figure is a square with a semicircle on one side. Find
a) the lengths of BC, AB and DC
b) the radius of the semicircle
c) the length of the arc BC
d) the perimeter of the figure.

4.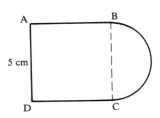

The figure is a square with a semicircle on one side. Find
a) the radius of the semicircle
b) the length of the arc BC
c) the perimeter of the figure.

5.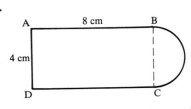

The figure is a rectangle with a semicircle on one end. Find
a) the length of BC
b) the radius of the semicircle
c) the length of the arc BC
d) the perimeter of the figure.

Circle Calculations 393

6.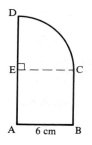

The figure is a square with a quadrant on one side. Find
a) the lengths of EC and AD
b) the length of the arc DC
c) the perimeter of the figure.

7.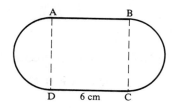

The figure is a square with semicircles on two ends. Find
a) the radius of each semicircle
b) the lengths of the arcs AD and BC
c) the perimeter of the figure.

8.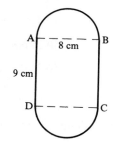

The figure is a rectangle with semicircles on two ends. Find
a) the radius of each semicircle
b) the lengths of the arcs AB and CD
c) the perimeter of the figure.

9.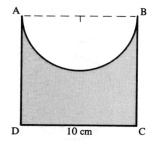

The figure is a square with a semicircle cut from one side. Find
a) the radius of the semicircle
b) the length of the arc AB
c) the perimeter of the figure.

THE AREA OF A CIRCLE

The formula for finding the area of a circle is

$$A = \pi r^2$$

If a circle is cut into sectors and the pieces placed together as shown, we get a "rectangle".

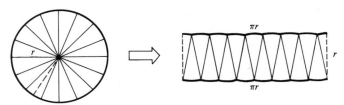

Area of circle = area of "rectangle".
$$= \pi r \times r = \pi r^2$$

EXERCISE 28d

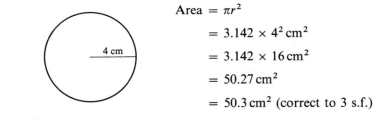

Take $\pi = 3.142$ and find the area of a circle of radius 4 cm.

Area $= \pi r^2$
$= 3.142 \times 4^2 \text{ cm}^2$
$= 3.142 \times 16 \text{ cm}^2$
$= 50.27 \text{ cm}^2$
$= 50.3 \text{ cm}^2$ (correct to 3 s.f.)

Take $\pi = 3.142$ and use your calculator to find the areas of the following circles.

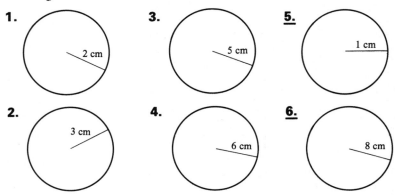

Circle Calculations 395

7. 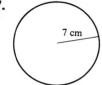 7 cm

10. 5.6 cm

13. 63 mm

8. 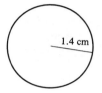 1.4 cm

11. 35 mm

14. 70 mm

9. 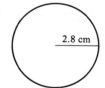 2.8 cm

12. 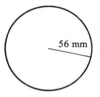 56 mm

15. 4.2 cm

EXERCISE 28e

Use $\pi = 3.142$ to find the area of a quadrant of a circle of radius 2 cm.

2 cm

(This is a quadrant, so its area is one quarter of the area of the circle.)

$$\text{Area of quadrant} = \tfrac{1}{4} \text{ of } \pi r^2$$
$$= \tfrac{1}{4} \times 3.142 \times 2^2 \text{ cm}^2$$
$$= \tfrac{1}{\cancel{4}} \times 3.142 \times \cancel{4} \text{ cm}^2$$
$$= 3.14 \text{ cm}^2 \text{ (correct to 3 s.f.)}$$

For this exercise use π = 3.142.

1.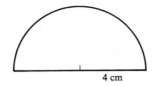

This is a semicircle. Its radius is 4 cm.
Find its area.

2.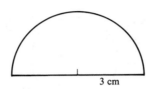

The radius of this semicircle is 3 cm.
Find its area.

3.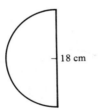

The *diameter* of this semicircle is 18 cm.
Find   a) its radius
        b) its area.

4.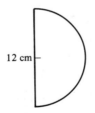

The diameter of this semicircle is 12 cm.
Find   a) its radius
        b) its area.

5.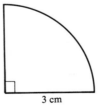

This is a quadrant. Find its area.

6.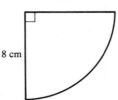

Find the area of this quadrant.

Circle Calculations   397

7.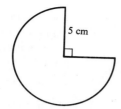

This is three quarters of a circle. Find its area.

8.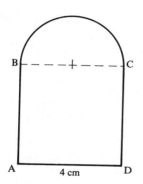

This figure is made up of a semicircle on one side of a square ABCD. The sides of the square are each 4 cm long. Find

a) the area of the square ABCD
b) the diameter of the semicircle
c) the radius of the semicircle
d) the area of the semicircle
e) the area of the whole figure.

9.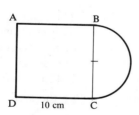

This figure is made up of a semicircle on one side of a square ABCD. The sides of the square are each 10 cm long. Find

a) the area of the square
b) the radius of the semicircle
c) the area of the semicircle
d) the area of the whole figure.

10.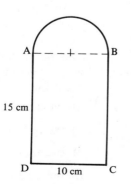

This figure is made up of a semicircle on one side of a rectangle which measures 15 cm by 10 cm. Find

a) the area of the rectangle
b) the radius of the semicircle
c) the area of the semicircle
d) the area of the whole figure.

11.

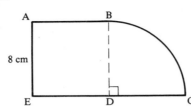

The diagram shows a square with a quadrant on one side. The sides of the square are each 8 cm long. Find

a) the area of the square

b) the radius of the quadrant

c) the area of the quadrant

d) the area of the whole figure.

12.

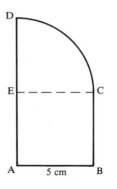

The diagram shows a quadrant on one side of a square. Find

a) the area of the square

b) the area of the quadrant

c) the area of the whole figure.

13.

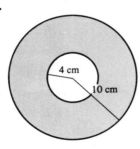

The diagram shows two circles with the same centre. The radius of the larger circle is 10 cm and the radius of the smaller circle is 4 cm. Find

a) the area of the larger circle

b) the area of the smaller circle

c) the shaded area between the two circles.

14.

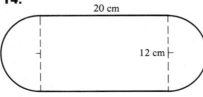

The diagram shows a rectangle with semicircles on the two short ends. Find

a) the area of the rectangle

b) the radius of each semicircle

c) the area of each semicircle

d) the area of the whole figure.

PROBLEMS

EXERCISE 28f Use $\pi = 3.142$ for this exercise and give your answers correct to 3 s.f.

1.

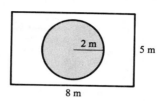

The diagram represents a rectangular lawn with a circular flower bed in the middle.

Find a) the area of the whole plot

b) the area of the flower bed

c) the area of the grassed part of the plot. (Use your answers to (a) and (b).)

d) the circumference of the flower bed

e) the number of bedding plants that can be placed round the edge of the flower bed if they have to be planted 0.5 m apart. (Use your answer to (d).)

2.

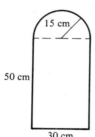

The diagram shows an arched window. The lower part is a rectangle measuring 30 cm by 50 cm. The upper part is a semicircle of radius 15 cm.

Find a) the total height of the window

b) the area of the rectangular part of the window

c) the area of the semicircular part of the window

d) the total area of the window correct to 3 s.f.

e) the cost of glass to glaze the window if glass costs 0.6 p per cm² (Use your answer to (d).)

f) the perimeter of the window.

3.

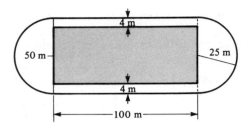

The diagram represents a sports ground. The shaded part is a rectangular grass pitch and the outer perimeter is a rectangle with semicircular ends.

Find a) the length and width of the grass pitch

b) the area of the grass pitch

c) the length of each semicircular arc

d) the perimeter of the ground

e) the number of people that can stand round the perimeter of the ground if the distance allowed for each person is 70 cm. (Use your answer to (d).)

4.

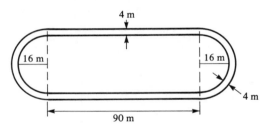

The diagram shows a running track which is 4 m wide. The straight sides are each 90 cm long. The radius of each *inner* semicircular end is 16 m.

Find a) the length of each *inner* semicircular arc

b) the perimeter of the *inside* of the track

c) the radius of each *outer* semicircular arc

d) the length of each *outer* semicircular arc

e) the perimeter of the *outside* of the track

f) the difference between the distance run by an athlete keeping to the outside of the track and the distance run by another athlete keeping to the inside of the track. (Use your answers to (b) and (e).)

29 PERCENTAGE PROBLEMS

REMINDERS

1. As a fraction in its lowest terms, 16% is $\frac{16}{100} = \frac{4}{25}$

2. As a percentage, $\frac{11}{50}$ is $\frac{11}{50} \times 100\% = 22\%$

3. As a percentage, 0.47 is $0.47 \times 100\% = 47\%$

4. If 55% of the families in a town own a car, then $(100 - 55)\%$ i.e. 45%, do not own a car

5. 20% of £550 is $\frac{20}{100} \times £550 = £110$

6. 24 cm expressed as a percentage of 200 cm is $\frac{24}{200} \times 100\% = 12\%$

7. If 500 is increased by 20% the new value is $(100 + 20)\%$ or 120%, of the old value

    i.e. the new value is $\frac{120}{100} \times 500 = 600$

8. If 600 is decreased by 35% the new value is $(100 - 35)\%$ or 65%, of the old value

    i.e. the new value is $\frac{65}{100} \times 600 = 390$

PROFIT AND LOSS PER CENT

A profit or gain results when the selling price (SP) of an article is greater than the cost price (CP). When the selling price is less than the cost price a loss is made. Gain or loss is normally given as a percentage of the *original* or *cost price*.

If a store makes a profit of 40% on an article which it bought for £50, the profit is $\frac{40}{100} \times £50 = £20$, and the selling price is £50 + £20 = £70.

Notice that the cost price is 100% of itself.

When 40% is added, the selling price is 140% of the cost price

i.e. $\frac{140}{100} \times £50 = £70$

EXERCISE 29a

> A book bought for £12 is sold to give a gain of £3.
> Find the gain per cent.
>
> $$\text{Gain per cent} = \frac{\text{Gain}}{\text{CP}} \times 100\%$$
> $$= \frac{£3}{£12} \times 100\%$$
> $$= 25\%$$

Find the gain per cent in the following cases.

1. CP £8, gain £4
2. CP £20, gain £5
3. CP £50, gain £10
4. CP 40 p, gain 18 p
5. CP £1.60, gain 56 p
6. CP £95, gain £57
7. CP £42, gain £2.10
8. CP £36, gain £54
9. CP £12, gain £4.08
10. CP £55, gain £8.25

> A dealer bought a car for £1000 and sold it for £900.
> Find his loss per cent.
>
> $$\text{Loss} = \text{CP} - \text{SP}$$
> $$= £1000 - £900$$
> $$= £100$$
>
> $$\text{Loss per cent} = \frac{\text{Loss}}{\text{CP}} \times 100\%$$
> $$= \frac{100}{1000} \times 100\%$$
> $$= 10\%$$

Find the loss per cent in the following cases.

11. CP £75, loss £22.50
12. CP 76 p, loss 38 p
13. CP £200, loss £80
14. CP 48 p, loss 12 p
15. CP £90, loss £58.50
16. CP £7, loss £1.61
17. CP £8, loss £6.72
18. CP £1.80, loss £1.53

> An article costing £40 is sold at a profit of 30%.
> Find the selling price.
>
> $$\text{Profit} = \frac{30}{100} \times £40$$
> $$= £12$$
> $$\text{Selling price} = £40 + £12$$
> $$= £52$$

In each of the following questions find the selling price.

19. CP £100, profit 25% **23.** CP 140 p, loss 45%

20. CP £300, loss 30% **24.** CP £175, loss 8%

21. CP £5, loss 17% **25.** CP £45, profit 62%

22. CP £18, profit 55% **26.** CP 50 p, profit 32%

27. I bought a car for £8000 and sold it at a loss of 32%. How much did I lose?

28. A shop keeper buys a box of articles for £140 and sells them at a profit of 35%. Find his profit.

29. A jeweller buys a gold ring for £250 and sells it at a profit of 70%. Find the selling price.

30. A box of chocolates bought for £1.50 is sold at a profit of 24%. Find the selling price.

31. John buys a record for £4.50 and sells it for £3.15. Find his loss per cent.

32. A dealer buys a picture for £3500 and sells it for £4900. Find her gain per cent.

33. A publisher estimates that 10 000 copies of a book will cost £40 000. If she sells them at £6.50 each find her profit.

34. A dealer allows $12\frac{1}{2}$% off the list price of a car when no car is offered in part-exchange. If the list price is calculated by adding 60% to the cost price, how much profit does the dealer make on a car costing him £8000?

VALUE ADDED TAX

Value added tax or VAT is a government tax which is added to most of the things we buy and also to many services. The rate of tax is decided from time to time by the Chancellor of the Exchequer.

EXERCISE 29b

Find the purchase price of a record marked £7 plus VAT at 15%.

Method 1

Value added tax is 15% of the marked price of £7

i.e. $\text{VAT} = \frac{15}{100} \times £7 = £1.05$

Purchase price = marked price + VAT

$= £7 + £1.05$

$= £8.05$

Method 2

Purchase price is (100% + 15%) = 115% of the marked price

i.e. Purchase price $= \frac{115}{100} \times$ marked price

$= \frac{115}{100} \times £7$

$= £8.05$

In questions 1 to 15 find the purchase price of the given article or service assuming that the rate of value added tax is 15%.

1. A pair of shoes costing £26 + VAT.
2. A light fitting costing £62 + VAT.
3. A football marked £12.40 + VAT.
4. A calculator marked £12.80 + VAT.
5. A bucket marked £2.40 + VAT.
6. A camera marked £160 + VAT.
7. Dinner for three at £3.80 per head + VAT.

Percentage Problems

8. Cleaning 20 m² of carpet at 46 p per square metre + VAT.
9. Servicing the car for £56.60 + VAT.
10. My telephone bill amounting to £48.40 + VAT.
11. The hire of a spray gun for 4 days costing £8.50 per day + VAT.
12. 4 tyres for the car costing £28 each + VAT.
13. A packet of chocolate biscuits marked 40 p + VAT.
14. 5 m of timber costing £1.20 per metre + VAT.
15. 120 dinner plates costing £2.55 each + VAT.
16. Find the purchase price of a hairdryer marked £18 + VAT at 20%.
17. Find the purchase price of a suite of furniture marked £1600 + VAT at 18%.
18. A long case clock is marked £320 + VAT. How much must I pay for it when the rate of value added tax is 15%? How much more will it cost if the rate rises to 20%?
19. The price of a car is £5800 + VAT. What must a customer pay for it if the VAT rate is 15%? How much more would he have to pay if the VAT rate were 20%?
20. The price of a chair is marked as £80 + VAT at 20%. If the VAT rate is reduced to 15% what is the reduction in the total price of the chair?

DISCOUNT

We are familiar with large notices in shops at sale time which declare "10% off all marked prices" or "50% off all dresses on this rail". These are examples of what we call *discount*.

Discount is often given to encourage prompt payment, e.g. a builder's merchant may offer a discount of 5% if bills are paid within 7 days, $2\frac{1}{2}$% if paid within 14 days, otherwise strictly net, i.e. the full amount must be paid.

For example, 5% discount on a bill of £100 is $\frac{5}{100} \times £100 = £5$
and $2\frac{1}{2}$% discount is half as much, i.e. £2.50.
So payment within 7 days reduces the bill to £95
and payment within 14 days reduces it to £97.50.

After 14 days the full £100 is charged.

EXERCISE 29c

In a sale a chair marked £120 is offered at a discount of 15%. Find the sale price.

Method 1

Discount of 15% on a marked price of £120 is

$$\frac{15}{100} \times £120 = £18$$

Therefore the sale price = £120 − £18
= £102

Method 2

If discount is 15%, the sale price, or discounted price, is (100% − 15%) = 85% of the marked price

i.e. the sale price of the chair is $\frac{85}{100} \times £120$

= £102

In questions 1 to 10 find the cash discounts when articles at the given prices are sold at the given percentage discounts.

	Article	Marked Price	Percentage Discount
1.	Record	£6	30%
2.	Dress	£39.50	50%
3.	Jumper	£24	60%
4.	Motorcycle	£1850	15%
5.	Carpet	£120	35%
6.	Christmas cards	80 p	20%
7.	Bedroom suite	£1500	25%
8.	Book	£12.50	40%
9.	Computer	£192	$12\frac{1}{2}$%
10.	Bathroom suite	£550	$2\frac{1}{2}$%

Percentage Problems

In questions 11 to 15 find the cash price when the article is sold at the given discount.

	Article	Marked Price	Discount
11.	Bicycle	£110	10%
12.	Video	£660	20%
13.	Refrigerator	£150	15%
14.	Hi-fi unit	£475	50%
15.	Car	£9500	$12\frac{1}{2}$%

16. In a sale a dining suite marked £640 is offered at a discount of 25%. Find the sale price.

17. In a sale a dress marked £60 is offered at a discount of 30%. Find the sale price.

18. A tradesman offers a discount of 5% if he is paid in cash. How much is saved by paying an account for £125 in cash?

19. The discounts offered by a builders' merchant to local tradesmen are:

5% for payment within 3 days

$2\frac{1}{2}$% for payment within 7 days

otherwise strictly net.

How much is saved by paying a bill for £720
a) within 7 days,   b) within 3 days?

20. The marked price of a fur coat is £2000. In a sale there is a discount of 20%. There is a further discount of 5% of the sale price if I pay cash. How much would the coat cost me if I pay cash?

21. An illustrated book is published at £18 but if bought in a bookshop before January 1st its price is reduced by £1.50.
The book is also offered by a book club at a discount of 8%.
Would it be better to buy it in a shop at the pre-January price or through the book club?

FINDING THE ORIGINAL QUANTITY

Sometimes we are given an increased or decreased quantity and we want to find the original quantity. Consider, for example, a business man who buys a tyre for his van at £46 which includes VAT at 15%. Since he can reclaim the VAT from the government he will want to know how much the tyre actually costs him.

EXERCISE 29d

> A coat is sold for £78. If this gives a profit of 30% find the cost price.
>
> Let the cost price be £x
>
> If there is a profit of 30%
>
> $$\text{SP} = \frac{130}{100} \text{ of the CP}$$
>
> i.e. $78 = \frac{130}{100} \times x$
>
> $100 \times 78 = 130 \times x$ (multiplying both sides by 100)
>
> $\therefore \quad \frac{100 \times 78}{130} = x$ (dividing both sides by 130)
>
> $x = 60$
>
> Therefore the cost price of the coat is £60.

In questions 1 to 4 find the cost price.

1. SP £18, profit 50%
2. SP £57.50, profit 25%
3. SP £10.40, profit 30%
4. SP 99 p, profit 65%

5. A shopkeeper makes a profit of 20% by selling an article for £60. Find the cost price.

6. A car sold at £1375 gives the dealer a profit of 25%. What did he pay for it?

7. The cost of an electric kettle is £25.30 including VAT at 15%. What was its price before the VAT was added?

8. After receiving a discount of 30% Janet pays £24.50 for a skirt. Find its original price.

9. I buy a book for £5.85 after it has been reduced by 35% in a sale. What was its original price?

10. A retailer pays the wholesaler £437 to settle an account. If he was allowed a discount of 5% how much was the original account?

Ann sells a record to Cheryl for £3 at a loss of 40%. What did Ann pay for the record?

If there is a loss of 40% and the CP is £x

then $$\text{SP} = \frac{60}{100} \times \text{CP}$$

i.e. $$3 = \frac{60}{100} \times x$$

$$3 \times 100 = 60 \times x$$

$$\overset{1}{\cancel{3}} \times \frac{\cancel{100}}{\cancel{60}_{2}} = x$$

$$x = 5$$

i.e. Ann bought the record for £5.

In questions 11 to 14 find the cost price.

11. SP £10, loss 50%
12. SP £150, loss 25%
13. SP £9.10, loss 30%
14. SP 252p, loss 55%

15. A car dealer makes a loss of 35% by selling a car for £910. What did he pay for it?

16. A retailer makes a loss of 15% by selling an article for £28.90. How much did she pay for it?

17. By selling a video recorder for £481 Joan makes a loss of 26%. How much did she pay for it?

18. George bought a set of tools and sold them at a loss of 32%. If he lost £9.12 how much did he pay for them?

PROBLEMS

EXERCISE 29e

1. Janet's weekly wage is expected to increase by 10% each year. At present she earns £150 per week. How much will she earn next year?

2. Dried fruit bought at £55 per 50 kg bag is sold at £1.43 per kilogram. Find the total selling price and the percentage profit.

3. A second hand car dealer buys a car for £4500 and sells it at a loss of 8%. How much does he lose?

4. Bicycles bought for £80 are sold for £108. Find the gain per cent.

5. A business man pays £174.80 for four new tyres for his van. The price includes VAT at 15% which he is able to reclaim. How much do the tyres actually cost him?

John's weekly wage increases by 5% each year. This year he earns £160 a week. How much will he be earning per week in two years time?

First increase is 5% of £160

$$= \frac{5}{100} \times £160$$
$$= £8$$

∴ his weekly wage in one year's time will be £168

Second increase is 5% of £168

$$= \frac{5}{100} \times £168$$
$$= £8.40$$

∴ his weekly wage in two year's time will be

£168 + £8.40 = £176.40

---

6. The population of a town increases by 10% each year. In 1982 the population is 8000. Find the size of the population in
   a) 1983   b) 1984.

7. The value of a car decreases by 8% each year. Its value when new was £5000. What is its value after two years?

8. The height of a tree increased by 5% each year. At the beginning of a two year period it was 20 m high. How high was it at the end?

MIXED EXERCISES

EXERCISE 29f

1. Find the gain per cent if an article costing 65 p is sold for £1.04.

2. Jill bought a computer for £126 and sold it at a loss of 30%. How much did she lose?

3. Harry bought a car seat for £42 and sold it at a loss of $12\frac{1}{2}\%$. How much did he get for it?

4. Find the cash price of a record album marked £18.60 plus VAT at 15%.

5. In a sale a department store offers a discount of 20 p in the pound. What percentage is this? If the price of an overcoat is reduced by £12.80 find its original price.

6. A food mixer costs £41.40 which includes VAT at 15%. What is its price excluding VAT? How much will it cost if the VAT rate is increased by 5%?

7. Georgina made a profit of 35% by selling a bicycle for £91.80. What must she sell it for to make a profit of 45%?

EXERCISE 29g In this exercise several alternative answers are given. For each question write down the letter that corresponds to the correct answer.

1. 40% of £160 is

   A £40 **B** £64 **C** £640 **D** £400

2. Written as a vulgar fraction 15% is

   A $\frac{1}{15}$ **B** $\frac{20}{3}$ **C** $\frac{3}{20}$ **D** $\frac{1}{5}$

3. 4 expressed as a percentage of 16 is

   A 4% **B** 50% **C** 16% **D** 25%

4. A shopkeeper buys an article for £44 and sells it to give a profit of 50%. The selling price is

   A £66 **B** £22 **C** £88 **D** £94

5. If 30% of the number of pupils in a school is 300, the number of pupils in the school is

   A 700 **B** 9000 **C** 1000 **D** 330

6. 40 cm as a percentage of 2 m is

   A 40%   B 20%   C 200%   D 4%

7. A radio bought for £40 and sold for £30 results in a loss of

   A 10%   B 30%   C $33\frac{1}{3}$%   D 25%

8. John bought a moped and sold it a year later at a loss of 50%. Which of the following statements *must* be true?

   A John paid £100 for the moped.

   B John sold the moped for half as much as he paid for it.

   C The moped got very rusty while John owned it.

   D John had to find another £50 to buy a new moped.

9. Mr Black had a weekly wage increase of 5%. Which of the following statements *must* be true?

   A Mr Black's wages increased.

   B Mr Black got £5 more each week.

   C Mr Black's wage, before the increase, was £100.

   D Mr Black's old wage was 95% of his new wage.

30 MULTIPLE CHOICE REVISION EXERCISES

In all the exercises in this chapter, each question is followed by several alternative answers. Write down the letter that corresponds to the correct answer.

EXERCISE 1   ARITHMETIC

1. The fraction of the figure that is shaded is

    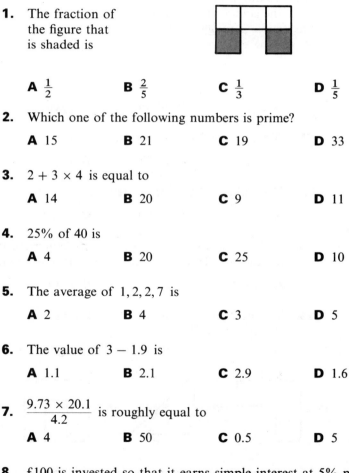

    A $\frac{1}{2}$ **B** $\frac{2}{5}$ **C** $\frac{1}{3}$ **D** $\frac{1}{5}$

2. Which one of the following numbers is prime?

    A 15   **B** 21   **C** 19   **D** 33

3. $2 + 3 \times 4$ is equal to

    A 14   **B** 20   **C** 9   **D** 11

4. 25% of 40 is

    A 4   **B** 20   **C** 25   **D** 10

5. The average of 1, 2, 2, 7 is

    A 2   **B** 4   **C** 3   **D** 5

6. The value of $3 - 1.9$ is

    A 1.1   **B** 2.1   **C** 2.9   **D** 1.6

7. $\dfrac{9.73 \times 20.1}{4.2}$ is roughly equal to

    A 4   **B** 50   **C** 0.5   **D** 5

8. £100 is invested so that it earns simple interest at 5% per annum. After 2 years the amount is

    A £110   **B** £10   **C** £5   **D** £120

9.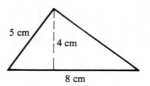

The area of the triangle is

A 32 cm²   **B** 20 cm²   **C** 16 cm²   **D** 40 cm²

10. Taking π as approximately 3, the circumference of the circle is roughly

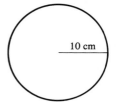

A 30 cm   **B** 60 cm   **C** 90 cm   **D** 15 cm

11. If 5 pens cost 50 p, 15 similar pens cost

A 10 p   **B** £1.50   **C** 15 p   **D** None of these

12. The ratio 2 : 5 can be written as

A 5 : 2   **B** 2.5 : 1   **C** 1 : 2.5   **D** 1 : 10

13. A shopkeeper gives a discount of 10% on marked prices. If he sells a radio for £18 what is the marked price of the radio?

A £19.80   **B** £20   **C** £180   **D** £1.80

14. The perimeter of one of the following figures cannot be found. Which one is it?

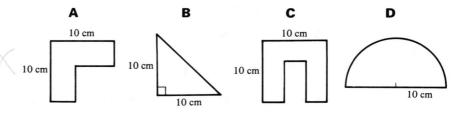

15. The profits of a company were shared between two partners in the ratio 2 : 3. The profits were £1500. The larger share was

A £1000   **B** £900   **C** £300   **D** £600

EXERCISE 2 GEOMETRY

1.

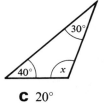

The angle marked x is

A 70°   **B** 110°   **C** 20°   **D** 290°

2.

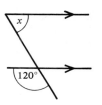

The angle marked x is

A 120°   **B** 30°   **C** 60°   **D** 90°

3.

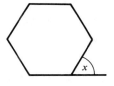

The figure is a regular hexagon. the angle marked x is

A 6°   **B** 120°   **C** 30°   **D** 60°

4. One of the following regular figures will not tessellate. Which one is it?

A   **C**

B   **D**

5. The number of lines of symmetry of an equilateral triangle is

A 3   **B** 6   **C** none   **D** 1

6.

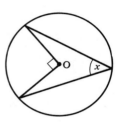

The angle marked x is

A 90°   **B** 45°   **C** 180°   **D** $22\frac{1}{2}°$

7.

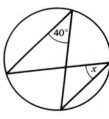

The angle marked x is

A 20°   **B** 50°   **C** 80°   **D** 40°

8. ABCDE is a pentagon. Which one of the following statements about ABCDE might *not* be true?

A The sum of the exterior angles is 360°.

B The figure has five sides.

C The figure has five lines of symmetry.

D The sum of the interior angles is 540°.

9.

Which one of the following three statements about △ABC is *not* true?

A The area of ABC is 12 cm²

B The length of AC is 5 cm

C AB² + BC² = AC²

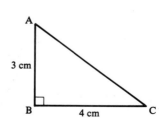

10.

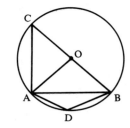

O is the centre of the circle. Which one of the following statements about this diagram must be correct?

A AOBD is a cyclic quadrilateral.
B $B\hat{A}C = 90°$
C $A\hat{D}B = 90°$
D $A\hat{O}C = A\hat{B}C$

EXERCISE 3 **ALGEBRA**

1. $x + 2y + x$ can be written as

    A $4x + y$ **B** $2x^2y$ **C** $4xy$ **D** $2x + 2y$

2. $2 - (x - 3)$ can be written as

    A $-x - 1$ **B** $5 - x$ **C** $2x - 6$ **D** $1 - x$

3. If $3x - 1 = 8$ then x is

    A $\frac{7}{3}$ **B** $\frac{1}{3}$ **C** 3   **D** 6

4. Given that $a = 2(b + c)$, the value of a when $b = 3$ and $c = 2$ is

    A 10   **B** 12   **C** 8   **D** $2\frac{1}{2}$

5. If $p = s - t$ then, when $s = 4$ and $t = -1$, p is

    A 3   **B** -4 **C** 5   **D** 4

6. $x(2x - y)$ can be written as

    A $3x - y$ **B** $2x^2 - xy$ **C** $-2x^2y$ **D** $2x^2 - y$

7. When $I = 2s + t$ is written with t as the subject, the formula is

    A $t = \frac{I}{2s}$ **B** $t = I - 2s$ **C** $t = I + 2s$ **D** $t = 2s - I$

418   ST(P) Mathematics 3B

8. The number term in the expansion of $(x-2)(x-4)$ is

  A 8   **B** -8 **C** -6 **D** 6

9. The x term in the expansion of $(x-2)(x-4)$ is

  A $8x$ **B** $-8x$ **C** $-6x$ **D** $6x$

10. The x term in the expansion of $(x-4)(x-4)$ is

  A x^2 **B** $-8x$ **C** $8x$ **D** $16x$

EXERCISE 4 GRAPHS, TRANSFORMATIONS AND MATRICES

1.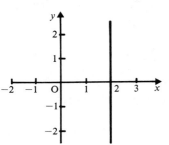

The equation of the line is

  A $x = 1$ **B** $y = 0$ **C** $x + y = 2$ **D** $x = 2$

2. The graph of the line $y = x - 1$ is

  A 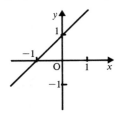   **B**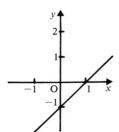

  C 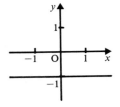   **D**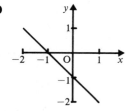

3.

The vector in the diagram is

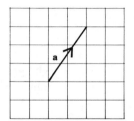

A $\begin{pmatrix} 3 \\ 2 \end{pmatrix}$ **B** $\begin{pmatrix} 2 \\ 3 \end{pmatrix}$ **C** $\begin{pmatrix} -2 \\ -3 \end{pmatrix}$ **D** $\begin{pmatrix} 2 \\ 2 \end{pmatrix}$

4.

After a rotation about O of 180°, the image of the triangle is

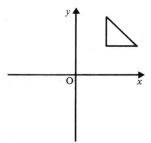

A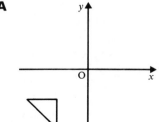

B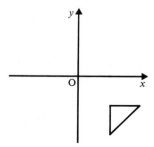

C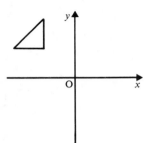

D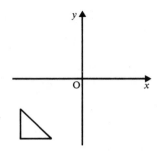

5.

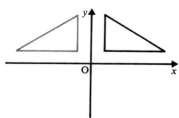

The grey triangle is the image of the black triangle produced by

A a translation parallel to the x axis

B an anticlockwise rotation about O of $90°$

C a reflection in the y axis

D a reflection in the x axis

6.

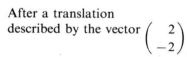

After a translation described by the vector $\begin{pmatrix} 2 \\ -2 \end{pmatrix}$ the image of P is the point

A $(4, 6)$ **B** $(4, 2)$ **C** $(0, 6)$ **D** $(2, 0)$

7. $\begin{pmatrix} 1 & 2 \\ 2 & 1 \end{pmatrix} + \begin{pmatrix} 0 & 3 \\ 1 & 1 \end{pmatrix}$ is equal to

A $\begin{pmatrix} 1 & 2 & 0 & 3 \\ 2 & 1 & 1 & 1 \end{pmatrix}$ **B** $\begin{pmatrix} 6 \\ 5 \end{pmatrix}$ **C** $\begin{pmatrix} 1 & 5 \\ 3 & 2 \end{pmatrix}$ **D** $\begin{pmatrix} 1 & 6 \\ 2 & 1 \end{pmatrix}$

8. $3\begin{pmatrix} 0 & 4 \\ 2 & 3 \end{pmatrix}$ is equal to

A $\begin{pmatrix} 0 & 12 \\ 6 & 9 \end{pmatrix}$ **B** $\begin{pmatrix} 3 & 7 \\ 5 & 6 \end{pmatrix}$ **C** $\begin{pmatrix} 3 & 12 \\ 6 & 9 \end{pmatrix}$ **D** $\begin{pmatrix} 1 & 2 \\ 6 & 9 \end{pmatrix}$

9. $\begin{pmatrix} 2 & 5 \\ 1 & 4 \end{pmatrix} - \begin{pmatrix} 1 & 9 \\ 2 & 4 \end{pmatrix}$ is equal to

A $\begin{pmatrix} 0 & 6 \\ 1 & 0 \end{pmatrix}$ **B** $\begin{pmatrix} 2 & 5 & -1 & -9 \\ 1 & 4 & -2 & -4 \end{pmatrix}$ **C** $\begin{pmatrix} 1 & -4 \\ -1 & 0 \end{pmatrix}$ **D** $\begin{pmatrix} 1 & 4 \\ 1 & 0 \end{pmatrix}$